595.771 DIS

BRITISH DIXIDAE
(MENISCUS MIDGES)
AND
THAUMALEIDAE
(TRICKLE MIDGES):
KEYS WITH ECOLOGICAL NOTES

by

R. H. L. DISNEY

Field Studies Council Honorary Research Fellow,
Department of Zoology, Cambridge University,
Cambridge

Illustrated by

C. JOAN WORTHINGTON AND THE AUTHOR

WITHDRAWN

FRESHWATER BIOLOGICAL ASSOCIATION
SCIENTIFIC PUBLICATION No. 56

1999

Series Editor: J. M. ELLIOTT

Sparsholt College Library
Sparsholt, Winchester
Hants. SO21 2NF

The Environment Agency welcomes the opportunity to part-sponsor the publication of *British Dixidae (meniscus midges) and Thaumaleidae (trickle midges): keys with ecological notes,* the latest in the FBA's series of Scientific Publications which continue to provide essential authoritative information needed to underpin the management of freshwater habitats.

The key and associated notes represent an invaluable source reference for identification, distribution and ecology. High quality information is required to maximise the effectiveness of measures to assess and monitor the state of rivers as part of the Environment Agency's aim to achieve a continuing improvement in the water environment in England and Wales. Moreover, accurate identification of rarer macroinvertebrates is essential with respect of the Agency's duty to further conservation.

The Environment Agency strongly recommends this FBA publication to practitioners and enthusiasts alike, but the Agency accepts no responsibility whatsoever for any errors or mis-statements contained in this publication. Any queries or comments should be directed to the FBA.

Published by the Freshwater Biological Association,
The Ferry House, Far Sawrey, Ambleside, Cumbria LA22 0LP

© Freshwater Biological Association 1999

ISBN 0 900386 60 6

ISSN 0367-1887

PREFACE

In 1975 the Freshwater Biological Association published a key to the larvae, pupae and adults of the British Dixidae. This became the standard work on this small but fascinating group of insects, sometimes known as the meniscus midges. The latter name refers to the larvae which are frequently found suspended from the surface of still or slow-flowing waters. They are therefore particularly sensitive to surface pollutants, and their sudden disappearance may be indicative of such pollution. In this new publication, the keys to Dixidae have been completely revised and the British list has been extended from fourteen to fifteen species. In addition, there are now keys to the larvae and adults of the three species in the small family Thaumaleidae, sometimes known as the trickle midges. The latter are often identified wrongly as chironomid larvae.

The original key stimulated interest in the Dixidae, as shown by the more extensive list of references in this new publication. As in all recent Scientific Publications of the FBA, there are sections on the ecology of both families. The last feature has proved to be a popular addition to these publications. This new publication should act as a stimulus to further work on the meniscus midges and should also introduce many to the largely neglected trickle midges.

We are most grateful to the Environment Agency for sponsoring this new publication, which will be of value to all those studying aquatic insects and will also introduce many to the largely neglected trickle midges.

The Ferry House
January 1999

J. M. Elliott
Series Editor

AUTHOR'S PREFACE

I became fascinated by the larvae of Dixidae when I was working on the blackfly vectors of River Blindness (Onchocerciasis) in West Africa. Indeed, I subsequently described the larvae of the two species I commonly encountered (Disney 1974a). On my return to Britain I found that half the British species occurred on the Malham Tarn Estate National Nature Reserve, which was then in my care. I decided, therefore, to attempt a key to the larvae of the British meniscus midges, by rearing larvae through to the adult stage. Furthermore, on looking through earlier collections of aquatic insects I had made in the vicinity of Flatford Mill in Suffolk, I found that I had collected a species new to the British List (Disney 1974b). By then I found myself embarked upon the first edition of this handbook, which represented my first significant taxonomic work as I began my shift from being primarily an applied ecologist (medical entomologist) to being primarily an insect taxonomist, who nevertheless remains at heart a naturalist and ecologist. However, it was during my time as a medical entomologist in Belize and Cameroon that I found that my ecological investigations frequently ended up in taxonomy (Disney 1994).

As I have prepared this second edition I have become only too aware of the deficiencies of the first edition, which has all the hallmarks of the self-taught novice taxonomist! However, I have learned that the publication of a new key frequently allows ecologists to progress in the study of individual, named species. This was certainly a result of the first edition of this work. Indeed, it is a gratifying feature of this new edition that I am able to cite a much more extensive list of publications referring to information on named species of Dixidae. I offer this new edition in the hope that it will not only facilitate further advance in our knowledge of British Dixidae but that the inclusion of the Thaumaleidae will act as a similar catalyst for an advance in our knowledge of this family also.

I have enjoyed producing it and hope that users will find it helpful in their enjoyment and study of two families whose larvae never cease to amaze and delight me.

Cambridge University Museum of Zoology — Henry Disney

CONTENTS

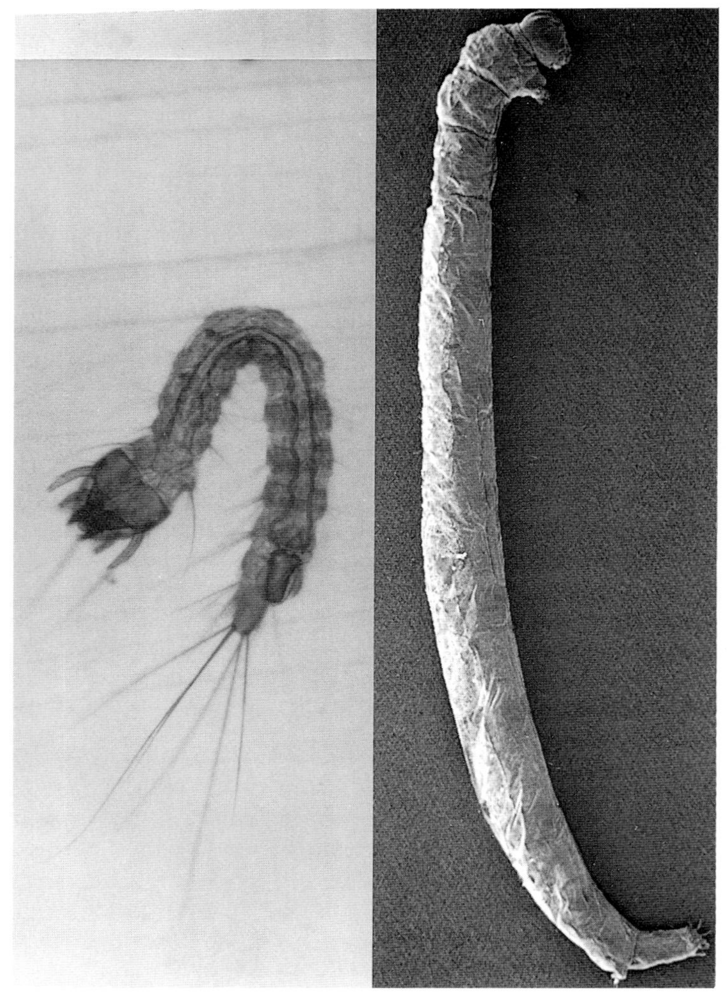

Plate 1. (*Above*): larva of *Dixella amphibia* (photograph of a live specimen, by E. K. Goldie-Smith). (*Below*): larva of *Thaumalea verralli* (SEM photograph of a preserved specimen, by R. H. L. Disney).

INTRODUCTION

The two families covered in this work are notable for their remarkable larvae (Plate 1) that are either associated with the meniscus on emergent structures, or occur in films of water at the margins of freshwater habitats. In order to carry out studies on named larvae, it is first necessary that someone rears every species from its larva to the adult stage. My own attempts to rear adults from the larvae of these two families, found on the Malham Tarn Estate National Nature Reserve in Yorkshire, were the origin of this work. Since then, adults of every species recorded from the British Isles have been reared from their larval stages.

The first edition of this work (Disney 1975) covered only the Dixidae (meniscus midges). This new edition updates the first, incorporating an addition to the British List (Disney 1992) and more recent ecological information. The distribution maps have been replaced by summary statements for each species. I have now added a treatment of the three British species of Thaumaleidae (trickle midges), whose larvae are not uncommon, but when encountered in the field are often not recognised as belonging to this family. It is hoped that by including these curious midges their neglect by freshwater biologists will be less excusable in the future. The possibility of additional species being found in Britain cannot be ruled out.

The larvae of meniscus midges and trickle midges are frequently encountered in unpolluted waters. While the former are not unfamiliar to most freshwater biologists, the larvae of trickle midges are often dismissed as odd-looking Chironomidae. By contrast, the adults of trickle midges are distinctive, although sometimes confused with blackflies (Simuliidae), whereas the adults of Dixidae are often mistaken for small, short-palped craneflies (Tipulidae). The larvae of both families may be commoner in running waters than those who restrict themselves to a particular sampling procedure may realise. The larger of the two British genera of meniscus midges (*Dixella*, with nine species) is characteristic of still waters. The smaller genus (*Dixa*, with six species) and the trickle midges (*Thaumalea*, with three species) are found in running waters, although the youngest and oldest larvae of the latter tend to avoid the flowing water itself and retreat to marginal microhabitats.

CLASSIFICATION

The Dixidae belong to the superfamily Culicoidea, in the order Diptera. By the end of the 19th century meniscus midges were recognised as a distinct family, even though fleas (order Siphonaptera) were still being treated as a family of Diptera (e.g. Theobald 1892). However, by the middle of the 20th century the Dixidae had been demoted to a subfamily of the Culicidae (e.g. Kloet & Hincks 1945), before advancing knowledge gave rise to restoration of the family status. More recent molecular information fully confirms the separate family status for these midges. Thus preliminary molecular studies on the relationship of the family to the Culicidae and Chaoboridae were presented by Rao & Rai (1990). The results indicate that the Chaoboridae is a sister group of the Culicidae, and the Dixidae is a sister group of the Chaoboridae. More recent studies of these and related families (Pawlowski *et al.* 1996) confirm these conclusions but place the Ceratopogonidae between the Dixidae and the Chaoboridae+Culicidae. They further propose the Simuliidae+Thaumaleidae as the sister group of the Dixidae, with the Chironomidae as the sister group of the former and as the basal group of the whole culicomorphan clade. However, a similar study by Miller, Crabtree & Savage (1997) concluded with two principal hypotheses. One places the Dixidae+Simuliidae as the sister group of the Ceratopogonidae at the base of the clade, with the Chironomidae+Chaoboridae+Culicidae as a subsequently evolved sister group of the Ceratopogonidae. Another hypothesis places the Dixidae as the basal taxon, with the Ceratopogonidae+Simuliidae and the rest of the families as two subsequently evolved clades. Their analysis did not include the Thaumaleidae.

A provisional phylogenetic reconstruction for the British species of Dixidae was presented by Disney (1983). Cladograms for the two genera are given in Fig. 1, that for *Dixella* being modified to include *D. graeca*. The clustering of this species with *D. filicornis* and *D. amphibia* is supported by subsequent observations on morphology of the eggs (see below, page 17).

Fig. 1. (*On facing page*). Cladograms for relationships between British species of *Dixa* (above) and *Dixella* (below). In the upper cladogram: PUB = *Dixa puberula,* NEB = nebulosa, DIL = *dilatata,* SUB = *submaculata,* NUB = *nubulipennis,* MAC = *maculata.* In the lower diagram: AES = *Dixella aestivalis*, AUT = *autumnalis*, SER = *serotina*, MAR = *martinii*, ATT = *attica*, FIL = *filicornis*, GRA = *graeca*, AMP = *amphibia*, OBS = *obscura*.

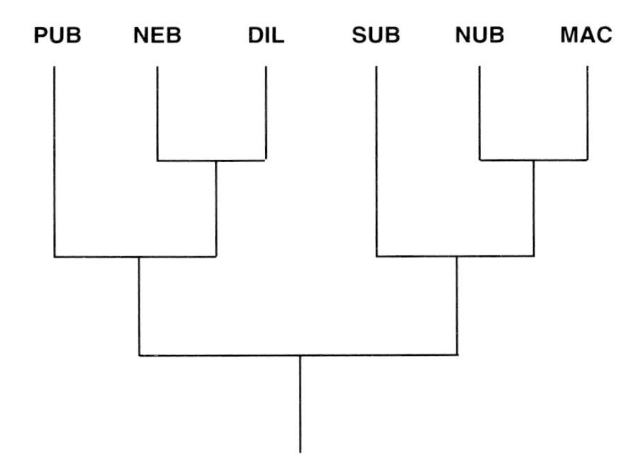
PUB
NEB
DIL
SUB
NUB
MAC

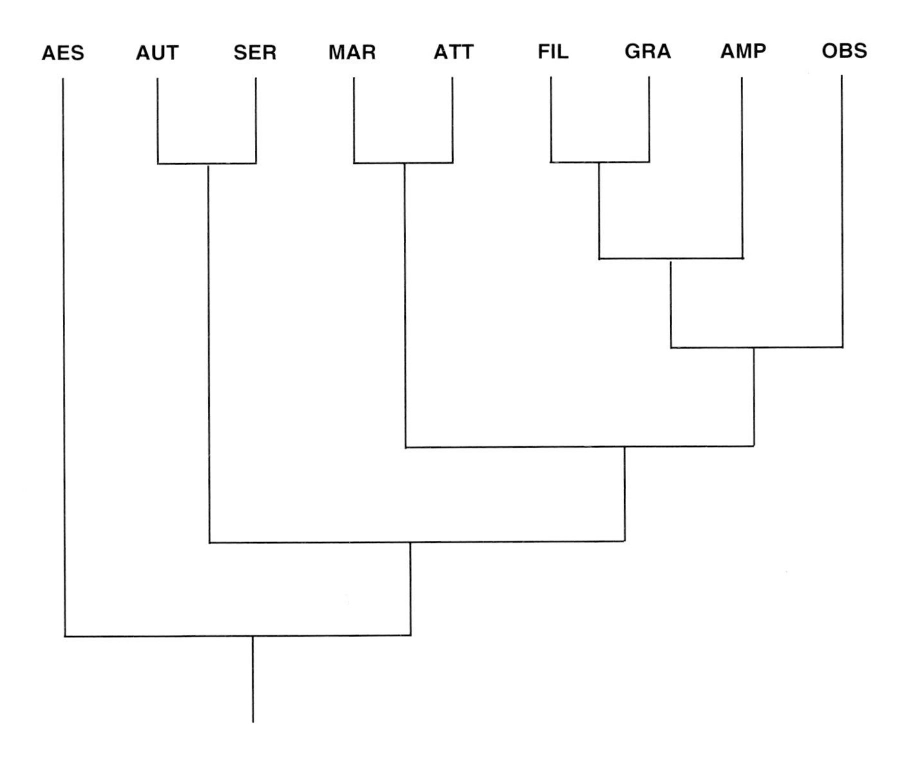
AES
AUT
SER
MAR
ATT
FIL
GRA
AMP
OBS

COLLECTING AND REARING

The eggs of Dixidae are typically laid in a mass of jelly on a solid substratum in the meniscus or just above, or else into streamside organic material. They can be placed in tubes of water from the habitat and kept in a cool shaded place to allow the larvae to hatch. The eggs of Thaumaleidae are laid individually amongst bryophytes or into cracks in rocks, and consequently are hard to find in the field.

The larvae of all Dixidae are characteristic of the meniscus on an emergent object, such as a sedge or a stone. They can also be found suspended from the surface film of still or slow-moving waters. They may temporarily submerge when disturbed. The larvae of *Dixella* are found in still waters, including regions of slack water in beds of emergent vegetation at the edges of rivers, but occasionally may be found at the margins of streams. The larvae of *Dixa* and *Thaumalea* are found in clean running waters, with the latter being characteristic of stony streams and springheads but *Dixa* being found in all types of stream. Larvae of *Thaumalea* are most commonly found in films of water flowing over rocks, but they may also be found in the meniscus on a leaf wrapped around a stone protruding above the water from the bed of a shallow stream. The smallest larvae tend to be hidden amongst bryophytes or in cracks in rocks. The oldest larvae tend to be in the damp zone out of the actual flowing water. *Dixa* larvae are usually more frequent in leaf packages. *Thaumalea* larvae move about in a film of water that is shallow enough to not fully submerge them. They readily frequent vertical rock surfaces, such as trickles running down a cliff face or an old wall. They prefer cold streams and are commonest in shaded positions.

Many larvae and pupae of Dixidae can be collected by direct examination of emergent structures. The pupae will normally be found just above the waterline. Frequently the last larval skin will be found beside the pupa. It should be collected and preserved in alcohol, and its association with the pupa kept for hatching should be noted. The pupae of *Thaumalea* can be found at the margins of their larval habitats in wet moss, amongst wet leaves or buried in mud. They may be at the bottom of streams. They can be extracted from samples of mud or stream-bottom debris with a small kitchen sieve gently sluiced by allowing the stream water to pass through the lower half of the sieve.

For general collecting of larvae and pupae of Dixidae from emergent vegetation, I use a bowl or similar receptacle. A plastic sandwich box is ideal. The receptacle is rapidly thrust into the water in such a manner that, as the water rushes in, it carries with it any larvae and pupae from the meniscus region. The receptacle is then held still and in a few moments the larvae will rise to the surface and start moving towards the meniscus at the sides.

For collecting larvae and pupae of both Dixidae and Thaumaleidae, emergent stones or leaf packets caught around stones can be washed off in a bowl of water. A white washing-up bowl is preferred for dislodging larval Thaumaleidae from stones. Where the stone is too large, or one is collecting from a film of water running over solid rock, then a paintbrush can be useful for picking off larvae. Disturbed *Thaumalea* larvae glide rapidly over a rock surface in a striking and characteristic manner that allows ready recognition in the field.

Adult meniscus midges may be found resting or on the wing. *Dixa* species commonly rest upside-down under low bridges, in culvert pipes, in spaces in drystone walls, under the leaves of streamside vegetation, under overhanging tree boles and similar situations. *Dixella* adults tend to rest on vertical surfaces with the head up and abdomen hanging down. Emergent or waterside sedges, reeds or rushes are favoured resting sites. The shorter-legged adult *Thaumalea* are most frequently found resting beside the larval habitat close to the water's edge. They rarely fly far from their breeding sites. In dry conditions, they can be collected with a sweep-net from grasses and ferns overhanging the banks of mountain streams. Otherwise they can be found sheltering under stones or leaves, from where they can be collected with a pooter (see below).

Adult Dixidae on the wing can be netted, especially those (such as *Dixa puberula*) that form loose swarms in the vicinity of water. Dry vegetation can be sampled with a sweep-net in order to procure resting midges. Individual resting meniscus midges or trickle midges may be collected by cautiously placing specimen tubes over them. They can also be collected with a pooter (aspirator). The model that I originally described in the first edition (Fig. 2) has become widely used for small insects. It consists of an approximately 40 cm length of narrower clear, polythene tubing (internal diameter 5 mm, external diameter 10 mm) one end of which is inserted into a 20 cm length of broader tubing (internal diameter 10 mm, external diameter 13 mm). The inserted end is first slightly narrowed by shaving its wall for about the first centimetre and then wrapping it with fine cotton gauze. After insertion, surplus gauze can be trimmed. Resting midges can be aspirated into the shorter tube and then blown into tubes of alcohol or dry tubes as required. It is useful to have several such pooters to allow for some getting too much condensation on their internal walls. These can be subsequently cleaned by dismantling the pooters and pushing wads of tissue through the tubes with a length of stout wire (such as is used for coat hangers).

Adult meniscus midges have been obtained in emergence traps set over the edge of a small loch (Morgan & Waddell 1961), at stream and pond margins (Roper 1962) and over a stream (Wagner 1980).

Adult Dixidae can be readily obtained by rearing from mature larvae and pupae. Individual larvae are put into specimen tubes half filled with water from the habitat. Specimen tubes with snap-on polythene caps have proved ideal.

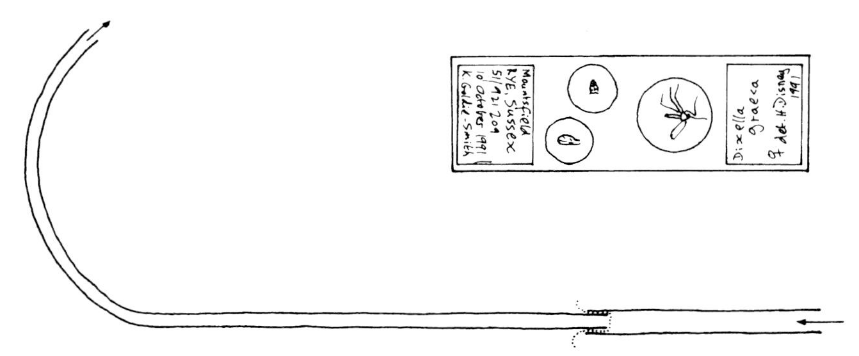

Fig. 2. A simple pooter (below) and a slide-mounted meniscus midge (above).

Four pinholes are made in the cap. The rearing tubes are stood in a shallow dish of water and placed in a cool place. The tubes are checked daily and the water replaced once a week. When a larva has pupated, the shed larval skin is rescued with a mounted needle or paintbrush and transferred to a labelled tube of alcohol, which is then attached to the side of the rearing tube with a rubber band or adhesive tape. When the adult emerges from the pupa, it is allowed to harden and darken for 24 hours before being added, along with its pupal skin, to the tube with the preserved larval skin. A larva may be reluctant to pupate unless an emergent sedge leaf or strip of dead sycamore leaf is added to the rearing tube. Goldie-Smith (1989a) gives more detailed instructions on laboratory rearing and also (1987) reports that some larvae may fail to complete their development because of a cytoplasmic polyhedrose virus infection that gives the larva a measled appearance (Plate 7, p. 83).

Living larvae of Thaumaleidae should be transported from the field to the laboratory in tubes that are damp, with water from their habitat, but without water slopping about, as the latter will tend to drown the larvae during transport. Pieces of vegetation, such as moss, can be added to the tubes.

I have succeeded in rearing trickle midges (with a success rate of approximately one adult from every three larvae) by isolating individual last-instar larvae in specimen tubes of at least 35 ml capacity, about one-third filled with water from their habitat. These tubes were placed in a cool shady place. Twice a day each tube was checked and the water replaced from a large jar of water from the habitat that had been kept in the light. Shed larval skins were carefully saved and preserved in alcohol. I have had somewhat less success rearing from penultimate-instar larvae. With younger larvae I have reared many through one moult, a few through two moults but none through to the adult stage.

Others have reported that the rearing of trickle midges has been achieved, with only partial success, by placing mature larvae in a Petri dish with scarcely enough water, from their habitat, to cover the bottom of the dish. The larvae of trickle midges are much more sensitive than Dixidae to becoming too warm. Leathers (1922) successfully reared larvae in a pocket, made from a double layer of cheesecloth, held within the mouth of one end of an old-lamp glass. A double-layered circle of cloth was secured over the top end of the glass in a manner that allowed the centre of the circle to be depressed. The larvae were placed in this inverted cone. A second double-layered circle of cloth was then secured, tightly stretched over the top of the cylinder. A ball of cotton wool was then placed in the centre of this cover and kept moist by a continuous slow drip of water. A modern adaptation of this system employs any wide glass or polythene cylinder (open at both ends) and circles cut from a pair of old stockings or tights instead of cheesecloth.

PRESERVATION AND MOUNTING

Specimens of all stages are best fixed and *preserved in 70% alcohol*. The most useful reference collections are those of slide-mounted specimens (see below for the preferred procedure).

There is a continuing debate about the most desirable *mounting medium* for small insects. Some (following the review of Upton 1993) place the highest priority on the longevity of the mount, while others emphasize the clarity of the cleared parts. Within the former preference some improve the clarity of the cleared parts by removing the soft tissues with a caustic agent (such as potassium hydroxide) before mounting in a medium such as Canada Balsam. Others, not wishing to sacrifice information that might be gleaned from examination of the soft tissues, use a mounting medium such as Euparal. My own preference is for *Berlese Fluid*. It, like Euparal, does not remove the soft tissues but changes their refractive index so that it approximates to that of glass. While it may not produce quite such a permanent mount as Euparal it tends to clear a little better, at least in the short term. The clearing of mounts in Euparal gradually improves over time.

Berlese Fluid can be purchased from various dealers. However, the formula has evolved over time and some dealers may use an inferior gum or mix in order to cut costs (but not the price!). It is advisable, therefore, to specify the formula when placing an order. The formula is as follows: gum arabic (picked lumps) 12 g, chloral hydrate crystals 20 g, glacial acetic acid 5 ml, 50% w/w glucose syrup 5 ml, distilled water 30-40 ml.

If preparing the mixture oneself, it is easiest to get rid of the impurities in the gum arabic by placing it in a clean cotton bag and immersing this in the distilled water until all the gum arabic has dissolved and passed through the bag into solution. The other ingredients are then added to this solution in the order shown above. A drop of the mixture is then tested by placing it on a slide to dry. If this drop does not dry to an amber-like texture, but is somewhat granular and whitish opaque, then the mixture contains too much sugar for the particular batch of gum arabic employed. The mixture should be diluted with an equal amount of a further mixture made up without the glucose syrup. If a tested drop of the resulting mixture is still not amber-like in appearance, dilute the mixture some more.

Making *slide mounts* is best done under a stereo (dissecting) microscope using fine (watchmaker's) forceps and mounted needles. To make a mount, the specimen is first briefly placed on a piece of tissue to remove the surplus alcohol. It is then placed in a drop of Berlese Fluid on a slide. After removal of a wing and the abdominal terminalia (see below) the parts are arranged in separate drops of Berlese Fluid and covered with coverslips (Fig. 2). It is best to use circular coverslips of 10 and 16 mm diameters.

In the case of *larvae*, most Dixidae are best oriented *dorsal side uppermost,* but Thaumaleidae are best placed with the *left side uppermost.* Six trickle midge larvae can be mounted under separate coverslips on one slide if 10 mm coverslips are used. Larger larvae can be cut in half, as the head and tail ends are the taxonomically significant parts.

With *pupae* or pupal skins, those of Dixidae are best mounted *on their sides* but those of Thaumaleidae should have the *dorsal face uppermost.* For Dixidae one first detaches the last few abdominal segments. The detached tail end is placed in a separate drop of Berlese Fluid and oriented with its caudal lobes as shown in Fig. 13B, C (p. 39)

With *adults*, the removal of the *right wing* is recommended, along with the *terminal segments of the abdomen* of both sexes for the Dixidae and for the males of the Thaumaleidae. With the British species of trickle midges the *abdomens of females* are best left attached to the thorax, for subsequent viewing from the left side. However, with specimens of many continental species, it is frequently necessary to view the ventral face of the abdominal terminalia. Therefore, where several specimens are available in a sample, some should be mounted in the same way as female Dixidae, in order to allow recognition of species not yet recorded from Britain. The wing is placed under one coverslip, the terminalia with the underside uppermost under another, and the rest of the midge with its left side uppermost under the third. In the case of the abdominal terminalia of a male dixid, this results in the dorsal side being uppermost, as post-emergence torsion rotates the terminalia through 180 degrees.

For *permanent storage*, slide-mounts in Berlese Fluid are dried on a hotplate until surplus medium at the edges of the coverslips shows small crazy-paving cracks. The coverslips should then be ringed. I have found that a modern "extra hard" nail varnish (such as those with the addition of nylon) is ideal for this purpose. Slides are best stored horizontally on trays housed in cabinets or storage boxes. In time, the mounts on vertically-stored slides may gradually creep down the slide.

EXAMINATION OF SPECIMENS

Many specimens can be identified by examining them in a watchglass of fluid (e.g. 70% alcohol) under a dissecting (stereo) microscope with a good spotlight, such as a fibre optic. For discernment of fine detail, specimens will need to be mounted on slides (see above) and transferred to a compound microscope. All the drawings in this handbook have been prepared from such slide-mounted specimens.

The morphological terms employed in the keys are intended to be those least likely to cause confusion. They make no attempt to be in line with current fashions. I endorse the views of authors such as Tuxen (1969), who wrote "the taxonomist should stick to the terms in common use for his group and no changes should be made for nomenclatural, morphological or semantic reasons". Thus the same part of the pleural region of the dixid adult thorax has a different label in the works of Nowell (1951), Christophers (1960), Peters & Cook (1966) or Oldroyd (1970). Likewise the same parts of the larval terminalia of Dixidae are differently labelled in the works of Johannsen (1934), Nowell (1951), Brindle (1963) or Peters & Adamski (1982). Similarly, the interpretation of the wing veins of Thaumaleidae poses a problem, as two relatively recent "standard" works of reference differ. In the Nearctic Manual, Stone & Peterson (1981) considered veins M1 and M2 to be separate. In the Australian textbook (Colless & McAlpine 1991), a single vein is interpreted as the M1+2, which seems to me more plausible; thus I agree with Sinclair (1996) in his revision of the eastern Nearctic species.

The interpretation of the homologies of the parts of insects is a continuing debate, with new information from developmental genetics, and other molecular data, increasingly requiring revisions of "established" interpretations. The temptation to keep revising the morphological terms in order to reflect these changes has caused considerable confusion for non-specialist users of taxonomic works. Thus the "sternopleuron" of an adult midge is also called the "katepisternum 2" or the "preepisternum 2" in different works. As long as it is accepted that the use of a term like sternopleuron does not necessarily imply homology with a part given the same label in a beetle or bee, such a traditional term is entirely acceptable. Those who contend otherwise do not follow their practice to its logical conclusion. Thus the use of the term "femur" has precedence in vertebrate anatomy long before its use in insect morphology; but not even the most extreme exponent of the view that every morphological label must only be employed for a unique set of homologous structures has suggested dropping terms such as femur and tibia from the literature of insect morphology!

DIXIDAE (Meniscus Midges)

The early literature on Dixidae is well summarised by Nowell (1951). Wagner (1997b) provides a brief update, noting that about 60 species are now known from the Palaearctic Region. The foundation work of Edwards (1920) formed the basis of Freeman's (1950) key to the adults of British species. However, the latter was based on pinned specimens, omitted two subsequently added species, and placed too much emphasis on small colour differences. The pioneering introduction to the larvae by Brindle (1963) omitted six British species. In preparing the first edition of this work (Disney 1975), I also consulted the works of Sicart (1959) and Vaillant (1959, 1965, 1969). Since then another species (*Dixella graeca*) has been added to the British List (Disney 1992). Some species, notably *Dixella autumnalis*, possess distinctive segregates in the adult stage. I have treated these as intraspecific varieties. Future work may require revisions of these opinions. Cytotaxonomy has been applied to Dixidae (Frizzi & Contini 1962; Frizzi, Contini & Mameli 1966). The use of mitochondrial DNA might be worth employing in future studies.

CHECKLIST OF BRITISH DIXIDAE

This list updates that of Kloet & Hincks (1976), including the addition of *D. graeca* (Disney 1992).

Genus *Dixa* Meigen, 1818

1. *dilatata* Strobl, 1900
    - *djurdjurensis* Vaillant, 1955
    - *riparia* Vaillant, 1959
2. *maculata* Meigen, 1818
    - *moesta* Haliday (in Curtis, 1832)
3. *nebulosa* Meigen, 1830
4. *nubilipennis* Curtis, 1832
5. *puberula* Loew, 1849
    - *sobrina* Peus, 1934
6. *submaculata* Edwards, 1920
    - *helenae* Sicart, 1958

Genus *Dixella* Dyar & Shannon, 1924
Paradixa Tonnoir, 1924
Dixina Enderlein, 1936
7. *aestivalis* (Meigen, 1818)
aprilina (Meigen, 1818)
cincta (Curtis, 1832)
8. *amphibia* (De Geer, 1776)
fuliginosa (Walker, in Curtis, 1832)
9. *attica* (Pandazis, 1933)
numidica (Sicart, 1955)
10. *autumnalis* (Meigen, 1838)
11. *filicornis* (Edwards, 1926)
inc. var. *immaculosa* Sicart, 1956
12. *graeca* (Pandazis, 1937)
obscura auctt. nec (Loew, 1849)
filicornis auctt. nec (Edwards, 1926)
13. *martinii* (Peus, 1934)
martini (Nowell, 1951)
laeta (Edwards, 1920) nec (Loew, 1849)
14. *obscura* (Loew, 1849)
15. *serotina* (Meigen, 1818)

Reference collections of slide-mounted British Dixidae have been deposited in the Natural History Museum in London, the Manchester Museum, the Zoology Museum of Cambridge University and the Cliffe Castle Museum in Keighley. The collection in Cambridge is the most comprehensive.

EGGS OF DIXIDAE

The eggs of Dixidae (Plates 2 and 3) are typically laid in a mass of grey or yellow-tinged jelly on a solid substratum or into streamside organic material. They have been described in detail by Goldie-Smith (1989a,b, 1990a, 1993) and Goldie-Smith & Thorpe (1991). Type I eggs are somewhat bulbous, with a reticulated surface microsculpture, and are found in *Dixella aestivalis, D. attica, D. autumnalis, D. martinii* and *D. obscura*. Type II eggs are more streamlined, with a smooth surface, and are found in *D. amphibia, D. filicornis* and *D. graeca*. Type III eggs are streamlined, with a spiculate surface microsculpture, and are characteristic of *Dixa* species.

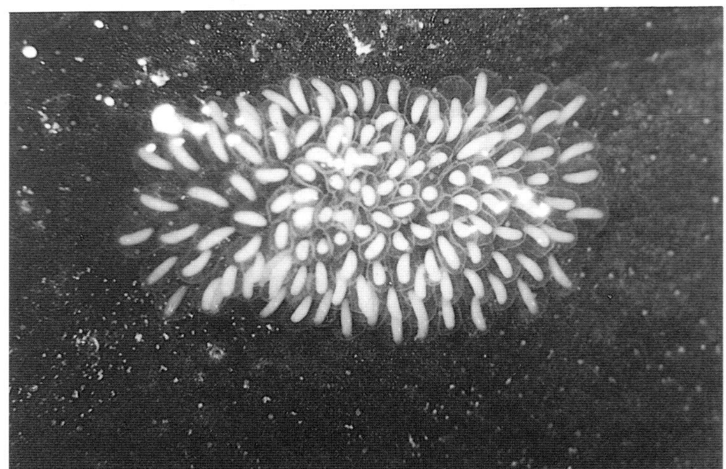

Plate 2. Egg-masses of Dixidae. Above: *Dixa nebulosa*. Below: *Dixella aestivalis*. (Photographs by E. K. Goldie-Smith).

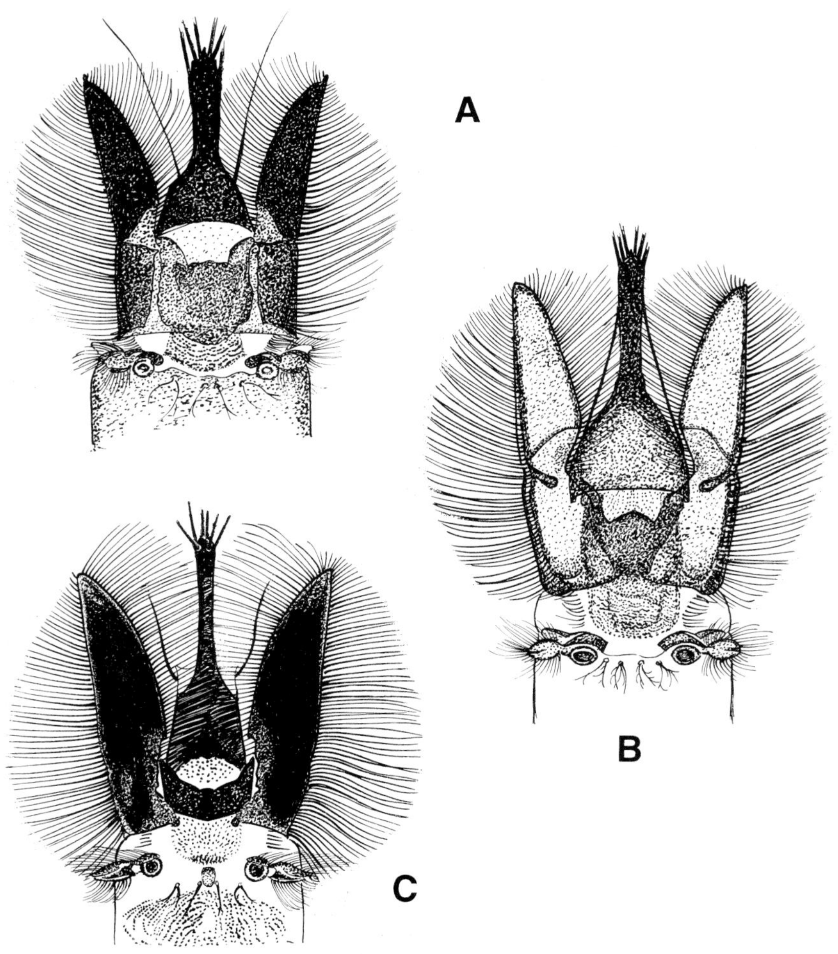

Fig. 5. Dorsal views of larval terminalia. **A**, *Dixa dilatata*; **B**, *D. nebulosa*; **C**, *D. puberula*.

4 Head (apart from dark hind margin), posterior paddles (apart from margins) and basal (anterior) part of caudal appendage are all mainly yellowish (Fig. 5B). In younger larvae these parts tend to be browner. The ventral row of denticles of each paddle, extending for a short distance from the tip along the inner margin, are all fine (Fig. 6A)—
Dixa nebulosa Meigen

— Head, paddles and caudal appendage uniformly dark brown (Fig. 5A). The ventral row of denticles of each paddle, extending for a short distance from the tip along the inner margin, are stronger and the terminal denticle is distinctly stronger still (Fig. 6B)—
Dixa dilatata Strobl

Fig. 6. (*On facing page*). Details of *Dixa* larvae. **A, B** – tips of paddles from below: **A**, *D. nebulosa*; **B**, *D. dilatata*. **C, D** – microtrichia on ventral face of head near paired setae behind maxillary palps: **C**, *D. nubilipennis;* **D**, *D. submaculata*. **E, F, G** – outlines of right paddles: **E**, *D. submaculata*; **F**, *D. maculata*; **G**, *D. nubulipennis*. (Scale bars = 0.1 mm).

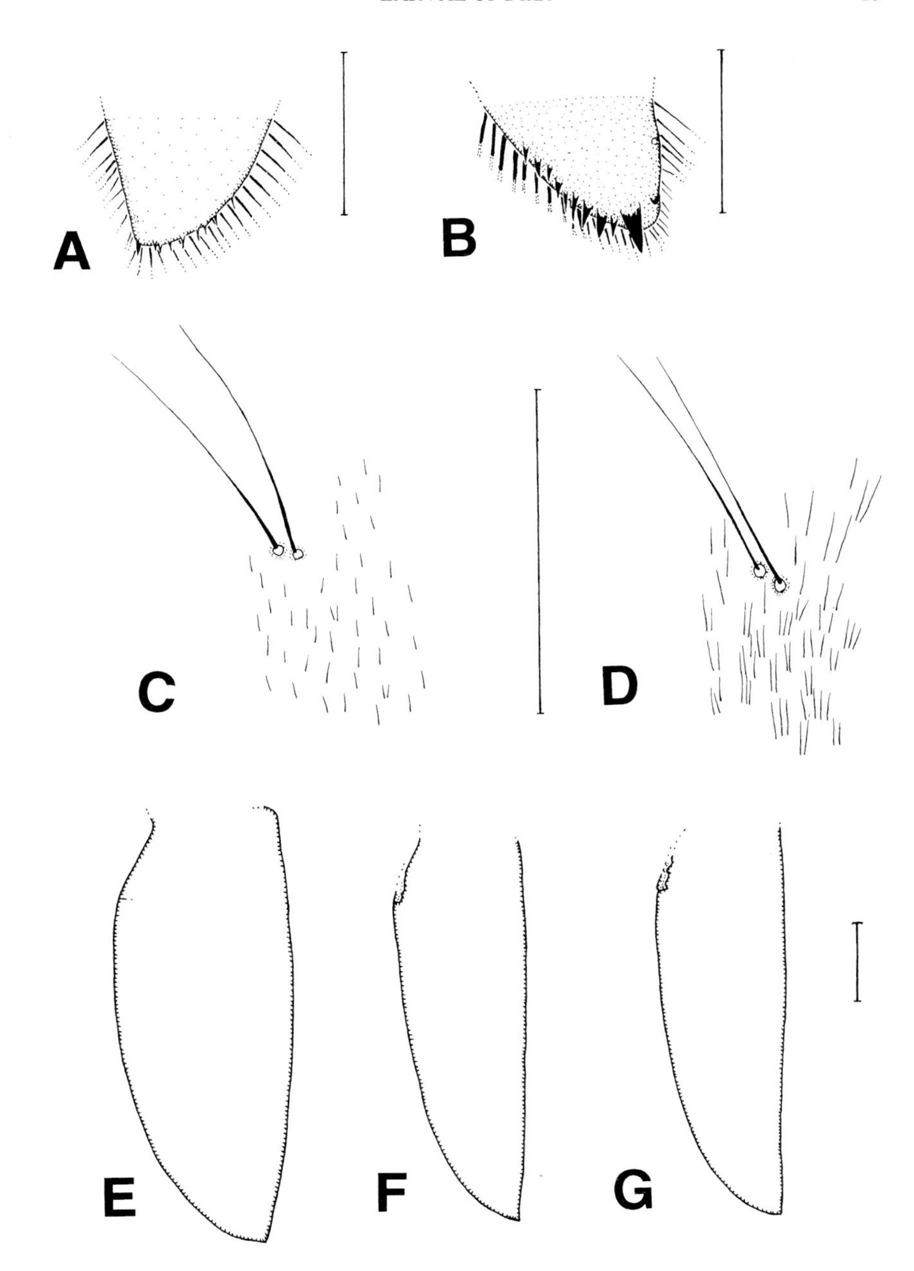
A
B
C
D
E
F
G

5 Posterior paddles relatively narrow (Figs 6F,G and 7A,C). Microtrichia (the minute hairs) adjacent to the pairs of fine setae on ventral face of head (situated directly behind the bases of the maxillary palps) *relatively short*, dark and mostly evenly spaced (e.g. Fig. 6C)—
Dixa maculata Meigen
and **Dixa nubilipennis** Curtis

[NOTE: so far no reliable means of separating the larvae of these two species has been discovered.]

— Posterior paddles relatively broad (Figs 6E and 7B). Microtrichia adjacent to the pairs of fine setae on ventral face of head *relatively long*, pale and unevenly spaced (Fig. 6D)— **Dixa submaculata** Edwards

Fig. 7. (*On facing page*). Dorsal views of larval terminalia. **A**, *Dixa maculata*; **B**, *D. submaculata*; **C**, *D. nubilipennis*.

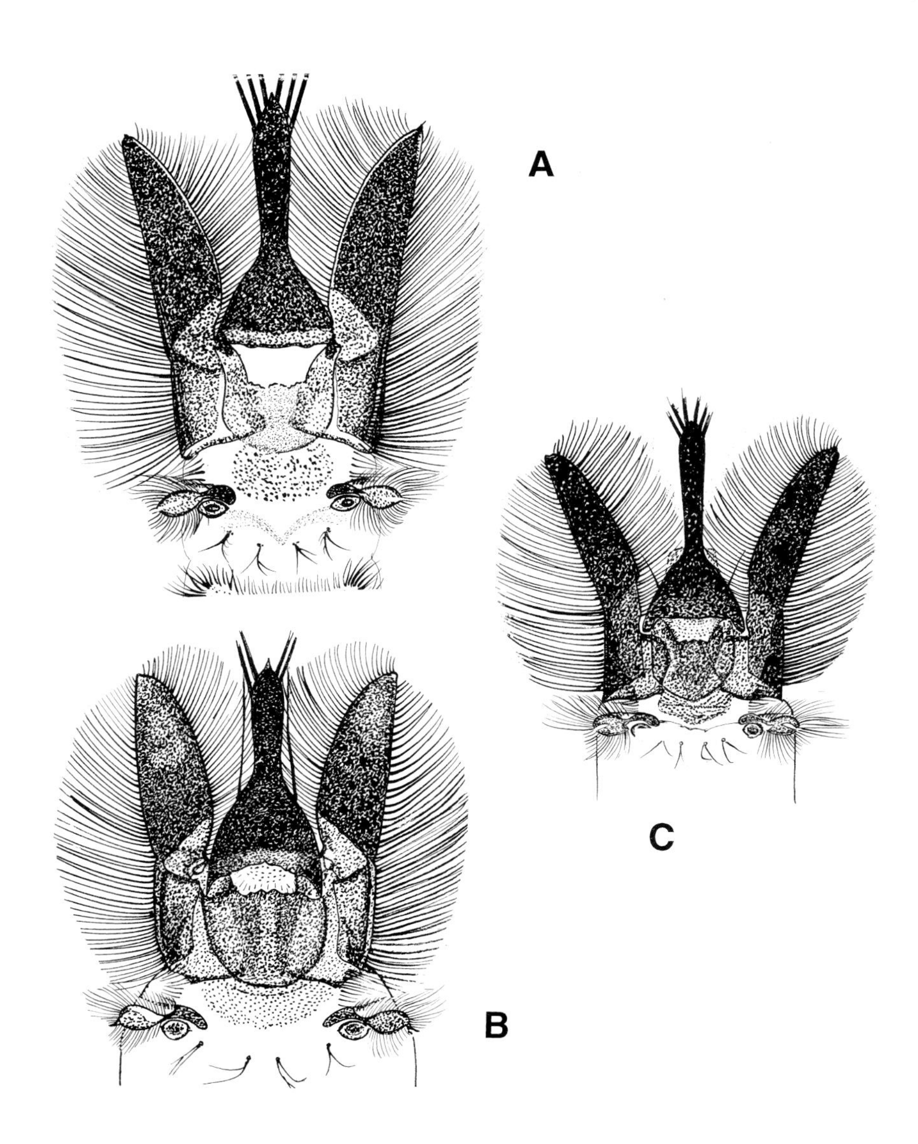
A
B
C

6 (2) The median posterior region of basal plate, whose hind margin is serrated or bordered with a fringe of tooth-like processes, projects rearwards (Figs 8↗ and 9E↗)— **7**

— The median posterior region of basal plate not or only slightly developed rearwards, so that the hind margin of the plate is essentially concave (Figs 10 and 12)— **9**

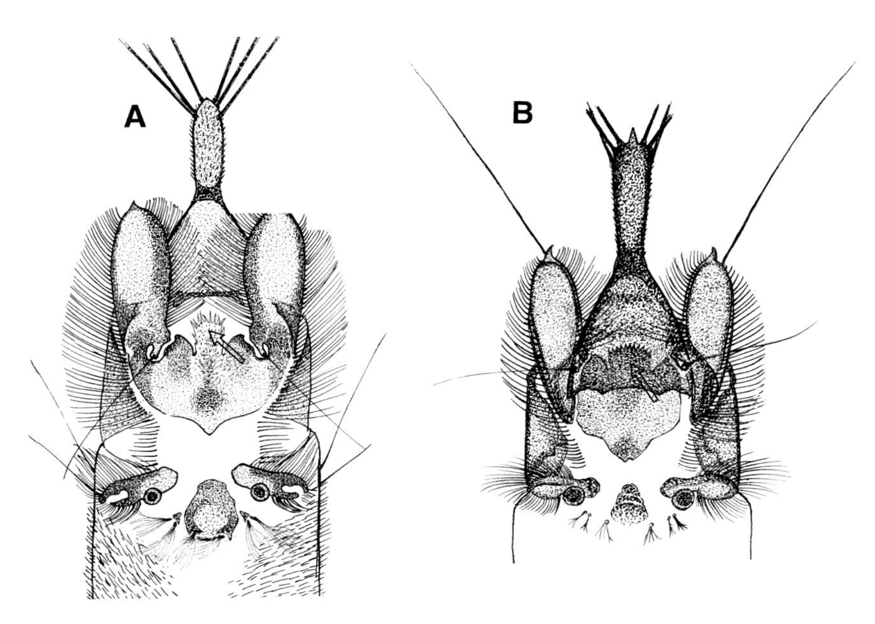

Fig. 8. Dorsal views of larval terminalia. **A**, *Dixella filicornis*; **B**, *D. amphibia*. ↗↗, posterior region of basal plates.

7 Median bars (MB, Fig. 4) on ventral combs (VC, Fig. 4) are simple, pale and somewhat irregular (Figs 11C, E). Ventral faces of paddles with small hairs in outer halves (Figs 9A, B). Body hair unusually long and dense (e.g. Fig. 8A)— **8**

— Median bars on ventral combs are distinctive T-shape (Fig. 11D). Ventral faces of paddles lack these fields of small hairs. Body hair shorter and less dense. Tail end of larva as in Fig. 8B— **Dixella amphibia** (De Geer)

8 The microtrichia (small hairs) on ventral face of paddle roughly equally spaced (Fig. 9B). Head capsule either bare (Fig. 9D) or with microscopic hairs only. Basal plate as in Fig. 9E— **Dixella graeca** (Pandazis)

— Many of the microtrichia on ventral face of paddle are in groups of 2-4 (Fig. 9A). Head capsule obviously hairy (Fig. 9C). Tail end of larva as in Fig. 8A— **Dixella filicornis** (Edwards

---

Fig. 9. (*On facing page*). Details of *Dixella* larvae. **A, B** – right paddles from below: **A**, *Dixella filicornis*; **B**, *D. graeca*. **C, D** – left sides of heads from below: **C**, *D. filicornis*; **D**, *D. graeca*. **E** – basal plate of *D. graeca* (↗ posterior margin).

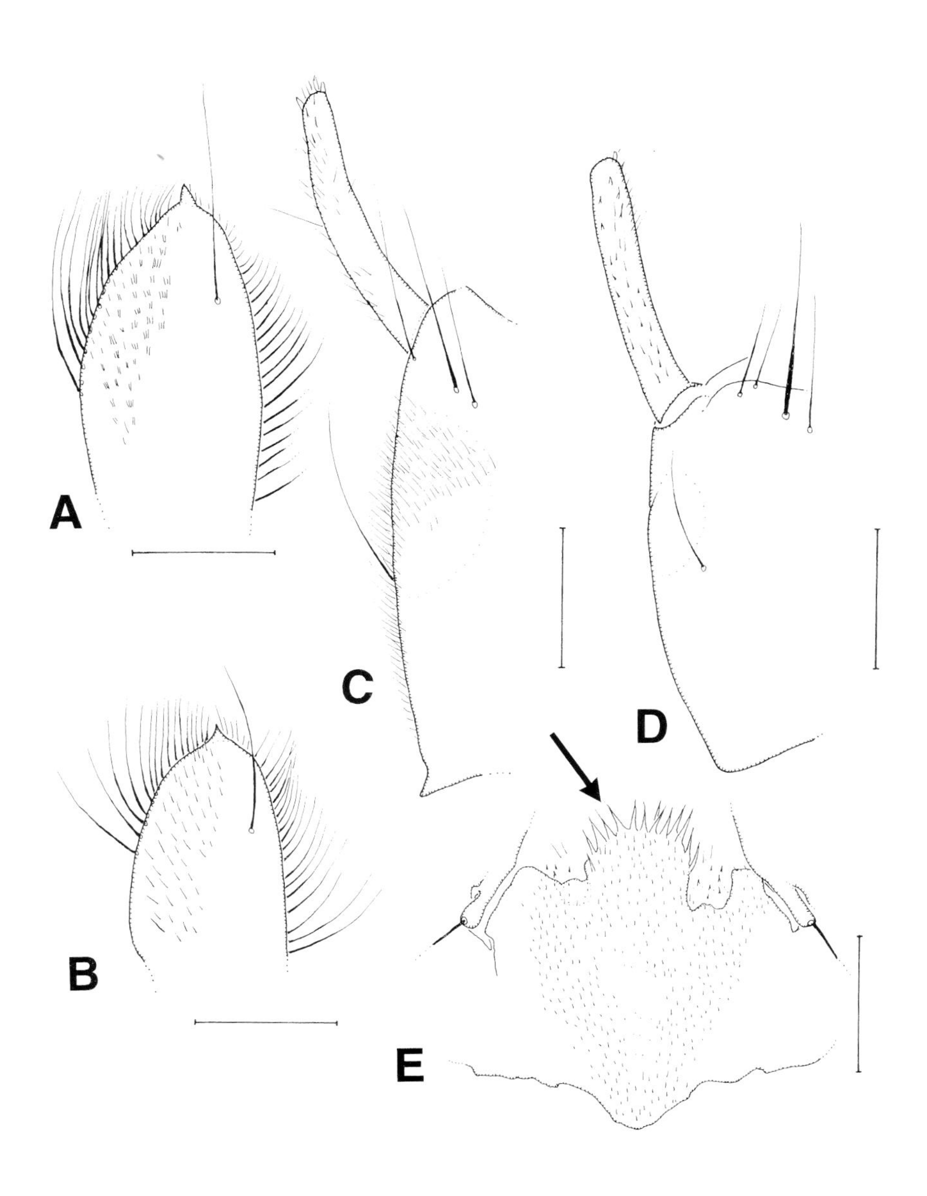
A
B
C
D
E

9 (6) Each pair of sub-spiracular ventral setae are closely approximated and arise from a single small platelet (Fig. 11H). Median bars of ventral combs as in Fig. 11F. Basal plate (BP, Fig. 3) distinctive in outline and a dark longitudinal stripe runs below lateral hairs of basal lobe of lateral plate (Fig. 10C ➚)— **Dixella aestivalis** (Meigen)

— Each pair of sub-spiracular setae clearly separate (e.g. Fig. 11I). Median bars of ventral combs otherwise (Figs 11A, B). Basal plates otherwise (Figs 10A, B and 12). Lateral plate dark stripe absent or obscure— **10**

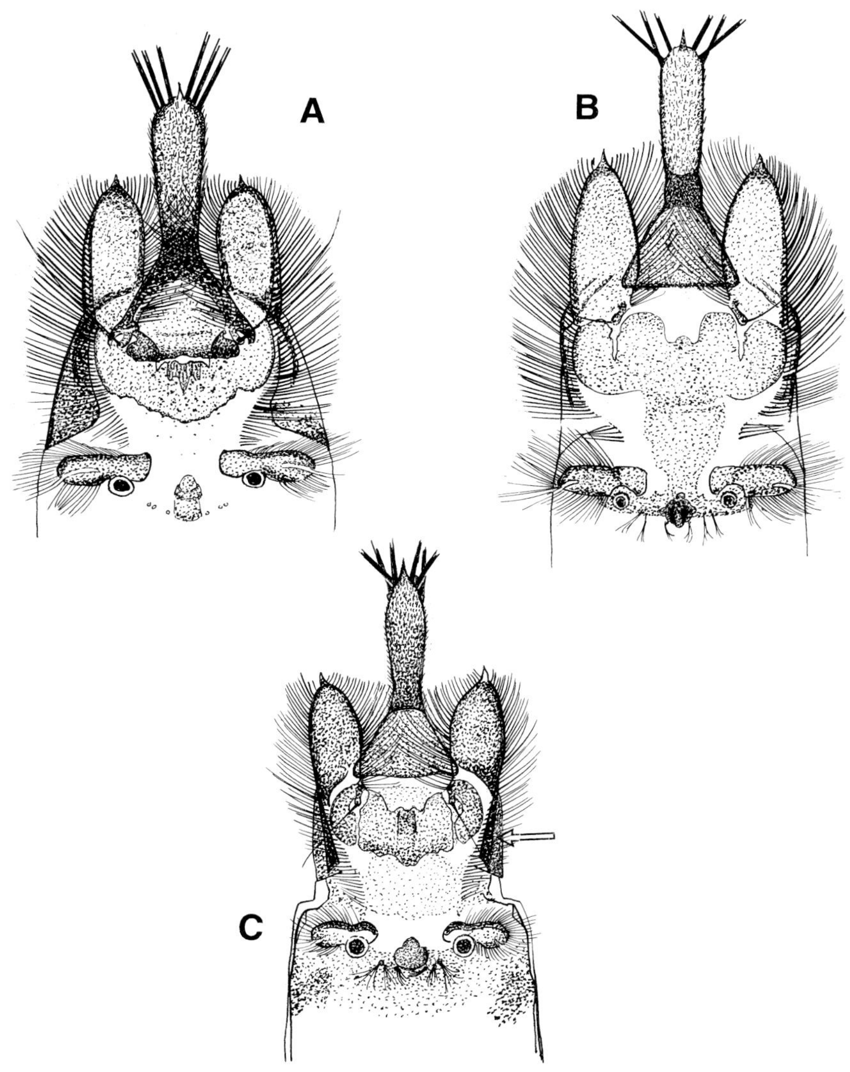

Fig. 10. Dorsal views of larval terminalia. **A**, *Dixella martinii*; **B**, *D. obscura*; **C**, *D. aestivalis* (↗ dark stripe).

10 Antennae brown to almost black. The dorsal median field of post-spiracular denticles, overlying the basal plate (see D and BP, Fig. 3), typically extends beyond the plate both anteriorly and posteriorly— **11**

— Antennae largely yellow, but a little darkened at tips. The post-spiracular denticles typically absent from anterior half of basal plate and from region in front of it— **12**

11 Inter-spiracular disc (ID, Fig. 3) as in Fig. 11G. Caudal appendage at most only a little darker than paddles. Tail end of larva as in Fig. 10B— **Dixella obscura** (Loew)

— Inter-spiracular disc as in Fig. 11K. At least middle third of caudal appendage clearly darker than paddles. Tail end of larva as in Fig. 10A— **Dixella martinii** (Peus)

Fig. 11. (*On facing page*). Details of *Dixella* larvae (for location of parts, see Fig. 4). **A-F** – median bars of anterior ventral combs: **A**, *D. obscura*; **B**, *D. serotina*; **C**, *D. graeca*; **D**, *D. amphibia*; **E**, *D. filicornis*; **F**, *D. aestivalis*. **G** – inter-spiracular disc of *D. obscura*. **H, I** – bases of sub-spiracular ventral setae: **H**, left side of *D. aestivalis*; **I**, right side of *D. attica*. **J** – bases of anal setae on right side of *D. attica*. **K** – inter-spiracular disc of *D. martinii*. **L, M** – antennae: **L**, *D. autumnalis*; **M**, *D. attica*. (Scale bars = 0.1 mm).

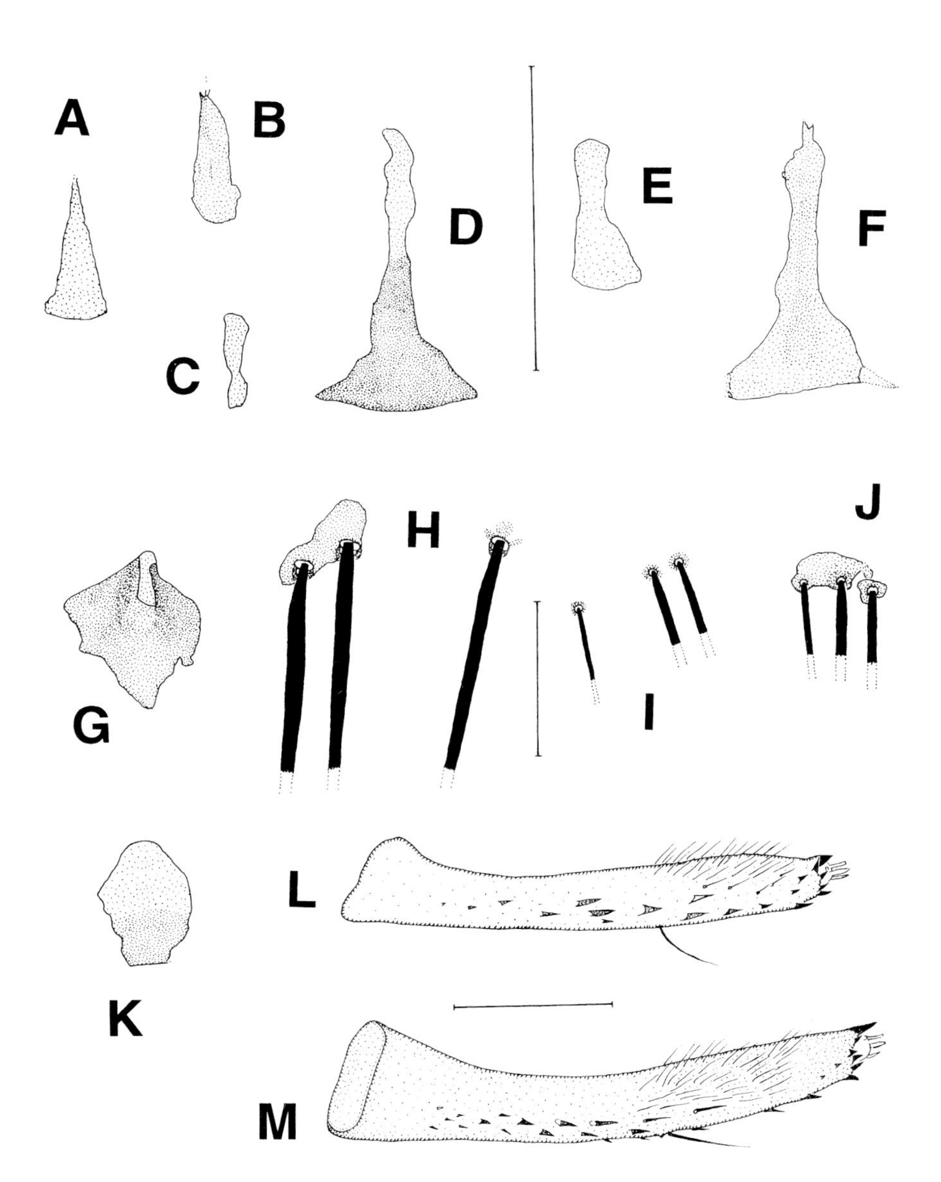
A
B
C
D
E
F
G
H
I
J
K
L
M

12 Posterior portion of caudal appendage (beyond the fracture line that separates it from the anteriorly-widening basal portion) is relatively narrow (compared with widths of paddles) (Fig. 12C) and its ventral face is almost uniformly covered in small spinous hairs—
Dixella serotina (Meigen)

— Posterior portion of caudal appendage relatively broad (Figs 12A↗, 12B↗) and its ventral face with small spinous hairs variably restricted to a narrow median band or completely devoid of such hairs— **13**

13 Antennae with fewer spinules along shaft (Fig. 11L). Three anal setae (AS, Fig. 4) typically without any fusion of their basal platelets, but occasionally a shadowy platelet is discernible below the integument. Small hairs of ventral face of posterior portion of caudal appendage frequently absent. Tail end of larva as in Fig. 12A—
Dixella autumnalis (Meigen)

— Antennae with more spinules along shaft (Fig. 11M). Typically at least two of each set of anal setae arise from a common platelet (Fig. 11J). Small hairs of ventral face of posterior portion of caudal appendage typically form a continuous median band. Tail end of larva as in Fig. 12B— **Dixella attica** (Pandazis)

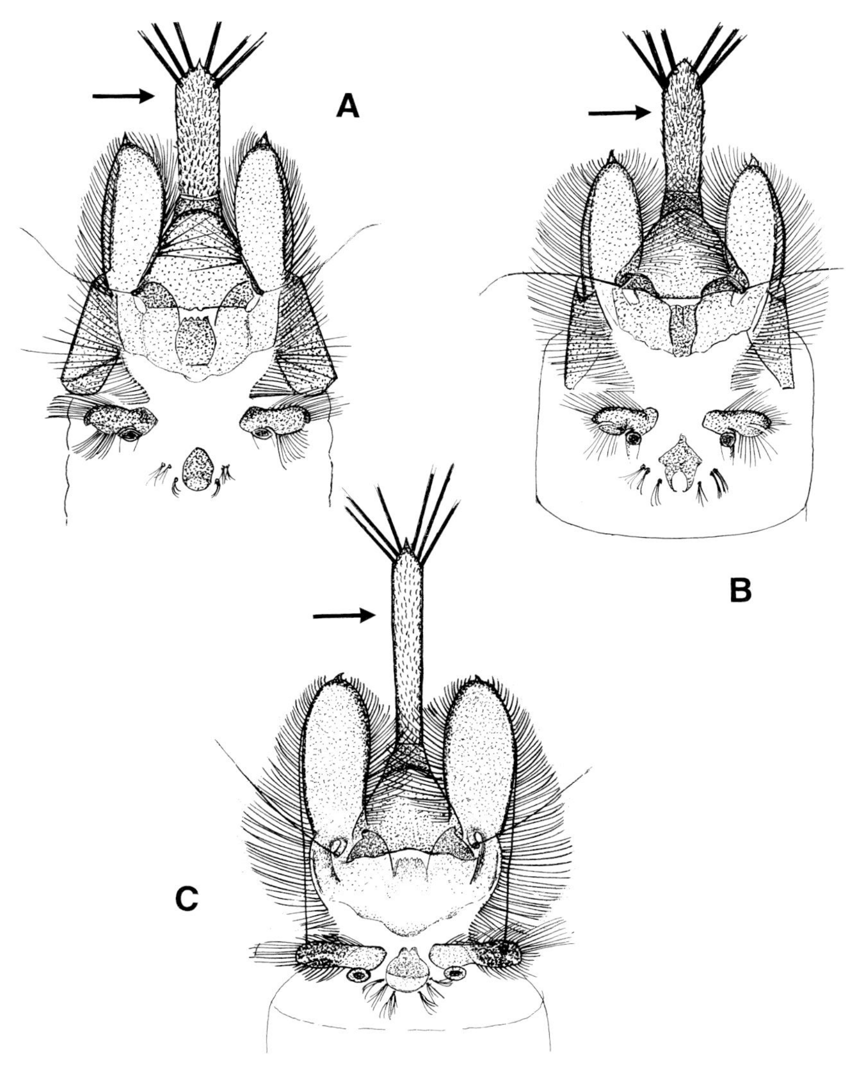

Fig. 12. Dorsal views of larval terminalia. **A**, *Dixella autumnalis*; **B**, *D. attica*; **C**, *D. serotina*. (↗ posterior portion of caudal appendage).

KEY TO PUPAE OF DIXIDAE

Pupae are frequently collected with their last larval skins still present beside their tail regions. The mounting of such larval pelts on slides will frequently provide more certain identifications than examination of the associated pupae. Mature pupae are best identified by hatching out the adult or, if already preserved, by dissection of the adult abdominal terminalia from the pupa. The following key may help with the recognition of the species.

1 Rim of respiratory trumpet with a deep, parallel-sided, emargination (Fig. 17amp)— **Dixella amphibia** (De Geer)

— Without this feature— **2**

2 Typically caudal lobes tapering gradually and the notch embraced by them is more U-shaped (Fig. 13C)— DIXA, **3**

— Typically caudal lobes more sharply tapered from a broader base and the notch embraced by them is more V-shaped (Fig. 13B)— DIXELLA, **8**

Fig. 13. (*On facing page*). **A**, *Dixa* pupa in side view: C, caudal lobes; M, mesonotum; R, respiratory trumpets; WS, wing sheaths. **B**, caudal lobes (ventral view) of a *Dixella* pupa; H, hypopygium. **C**, caudal lobes (ventral view) of a *Dixa* pupa: H, hypopygium.

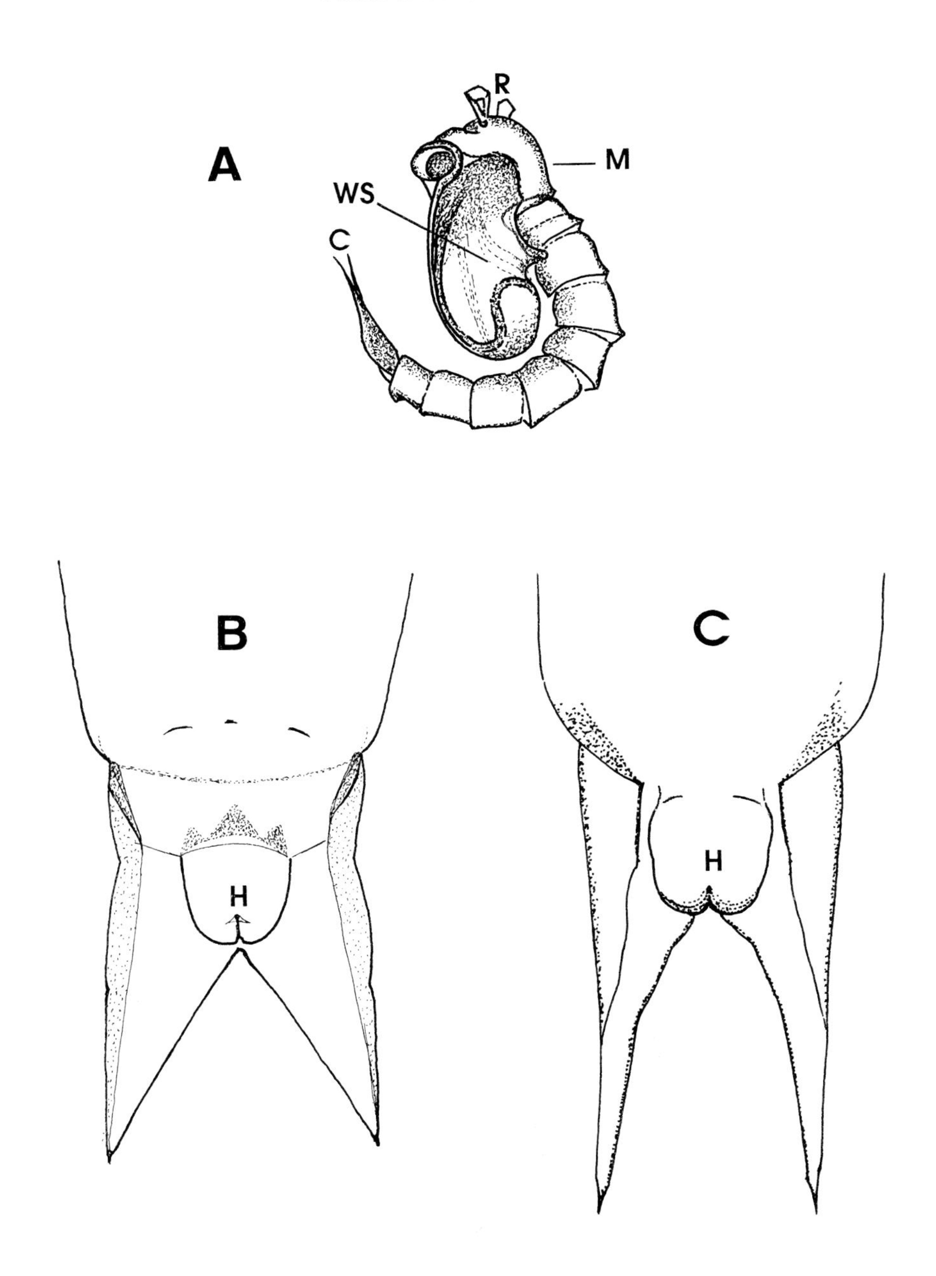
A
R
M
WS
C
B
H
C
H

3 Tip of caudal lobe suddenly narrows at position of conspicuous pre-apical tooth on outer margin (Fig. 14B ➚); otherwise marginal teeth are sparse (Fig. 14B). Respiratory trumpet as in Fig. 15sub—
Dixa submaculata Edwards

— Tip of caudal lobe more evenly tapered or, if suddenly narrowing, it is at level of a tooth on inner margin (Fig. 14E ➚) and the marginal teeth more numerous (Figs 14A, C-F). Respiratory trumpet otherwise (Fig. 15)— **4**

Fig. 13. (*On facing page*). Pupae of *Dixa*: tips of left caudal lobes from above. **A**, *Dixa dilatata*; **B**, *D. submaculata* (➚ narrow below tooth on outer margin); **C**, *D. nubulipennis*; **D**, *D. puberula*; **E**, *D. maculata* (➚ narrow below tooth on inner margin); **F**, *D. nebulosa*. (Scale bar = 0.1 mm).

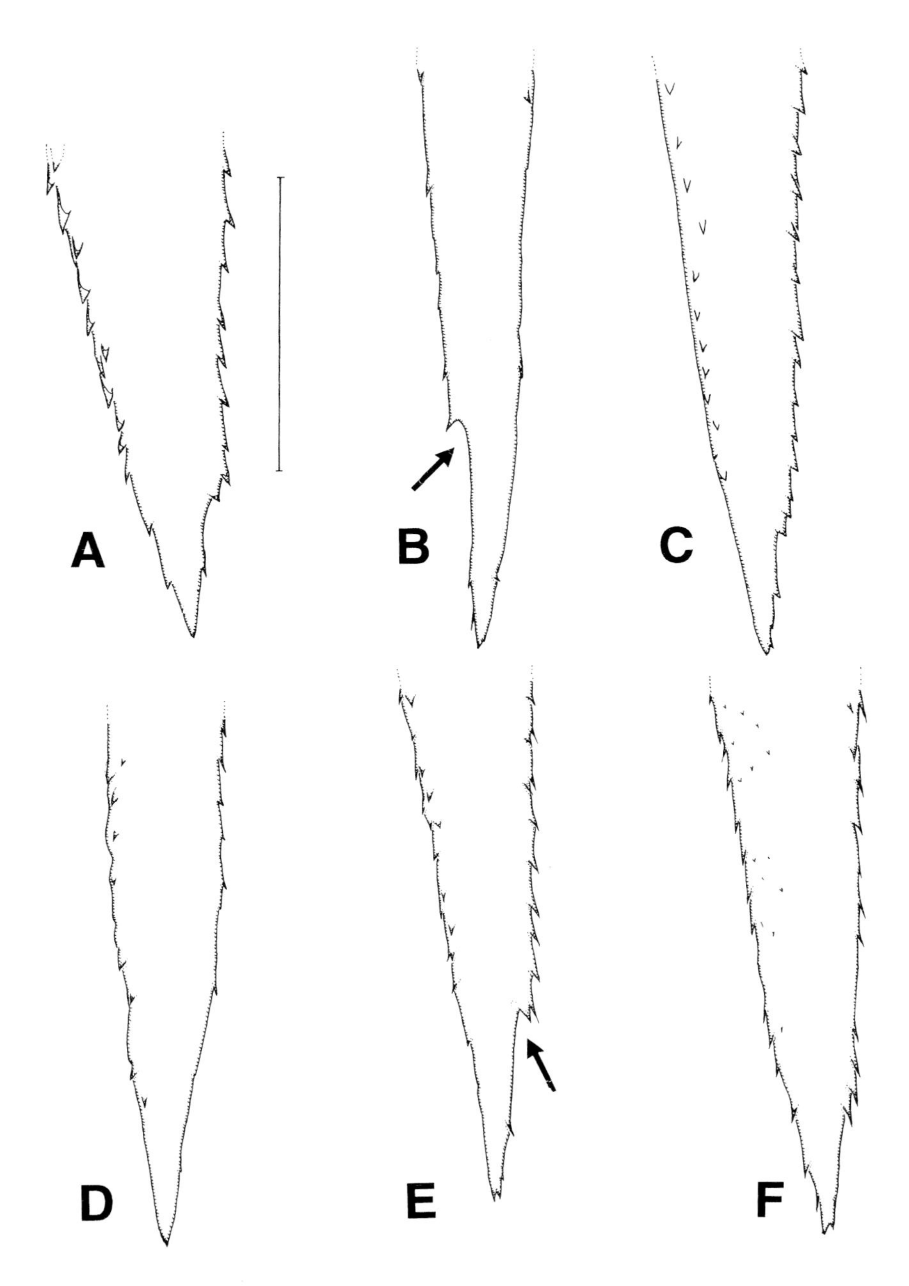
A
B
C
D
E
F

4 Respiratory trumpet a more widely flaring funnel, being about twice as wide at the rim as at the base (Fig. 15dil, nub)— **5**

— Respiratory trumpet not so widely flaring and width of rim only about 1.5 times that of base (Fig. 15pub, neb, mac)— **6**

5 Teeth on margins of each caudal lobe well developed in distal half (Fig. 14A). Respiratory trumpet as in Fig. 15dil— **Dixa dilatata** Strobl

— Teeth on margins of caudal lobe weaker, especially on outer margin (Fig. 14C). Respiratory trumpet as in Fig. 15nub— **Dixa nubilipennis** Curtis

6 Rim of respiratory trumpet clearly darker than rest of trumpet (Fig. 15pub). Caudal lobes with reduced teeth (Fig. 14D)— **Dixa puberula** Loew

— Rim of respiratory trumpet not obviously darker than rest of trumpet. Caudal lobes with more numerous and more conspicuous teeth (Figs 14E, F)— **7**

7 Caudal lobes as in Fig. 14E. Respiratory trumpet as in Fig. 15mac— **Dixa maculata** Meigen

— Caudal lobes as in Fig. 14F. Respiratory trumpet as in Fig. 15neb— **Dixa nebulosa** Meigen

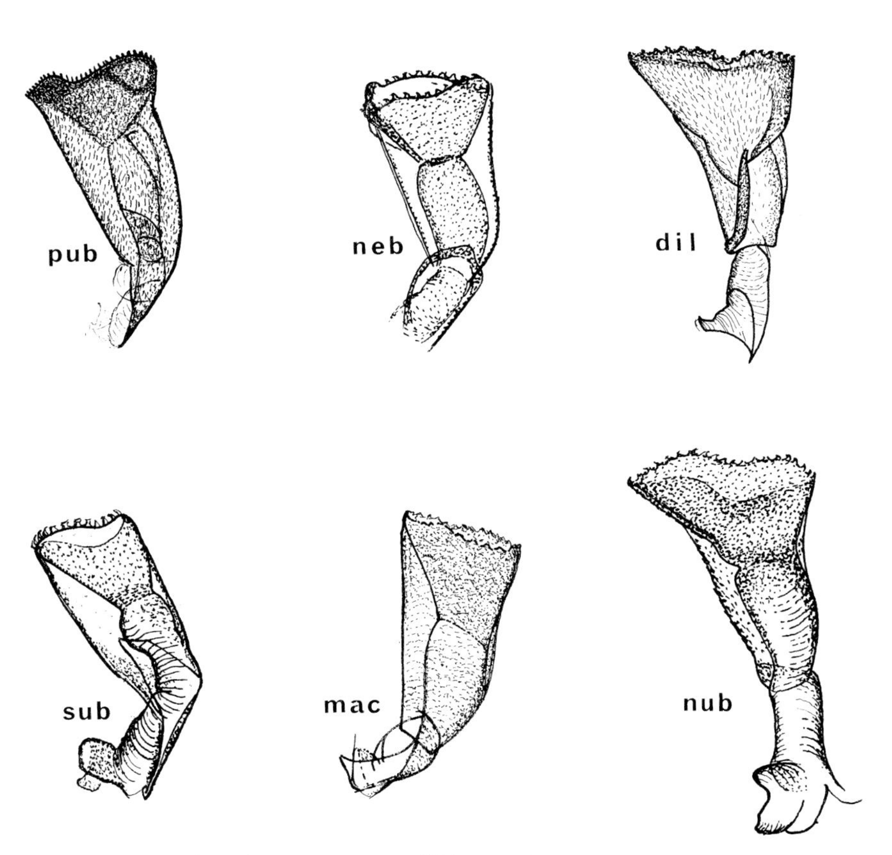

Fig. 15. Pupal respiratory trumpets of *Dixa* species.
Top row: pub, *D. puberula*; neb, *D. nebulosa*; dil, *D. dilatata*.
Bottom row: sub, *D. submaculata*; mac, *D. maculata*; nub, *D. nubulipennis*.

8 (2) Wing sheaths with fine denticles, similar to but weaker than those on mesonotum (the domed top of thorax above bases of wing sheaths; WS and M, Fig. 13)— **9**

— Wing sheaths bare— **10**

9 The denticles on side of upper half of first abdominal segment (in position of the 'a' in Fig. 16A) are linked by a fine network of ridges (Fig. 16B) and a similar network is found on metanotum (above the 'm' in Fig. 16A). Respiratory trumpet as in Fig. 17aut—
Dixella autumnalis (Meigen)

— The network of ridges on metanotum generally much weaker than on first abdominal segment. Respiratory trumpet as in Fig. 17ser—
Dixella serotina (Meigen)

Fig. 16. (*On facing page*). Details of *Dixella* pupae. **A** – left side of metathorax (m) and upper half of first abdominal segment (a); **B** – denticles and ridges on upper part of first abdominal segment of *D. autumnalis*; **C, D** – respiratory trumpets: **C**, *D. graeca*; **D**, *D. filicornis*. (Scale bars = 0.1 mm).

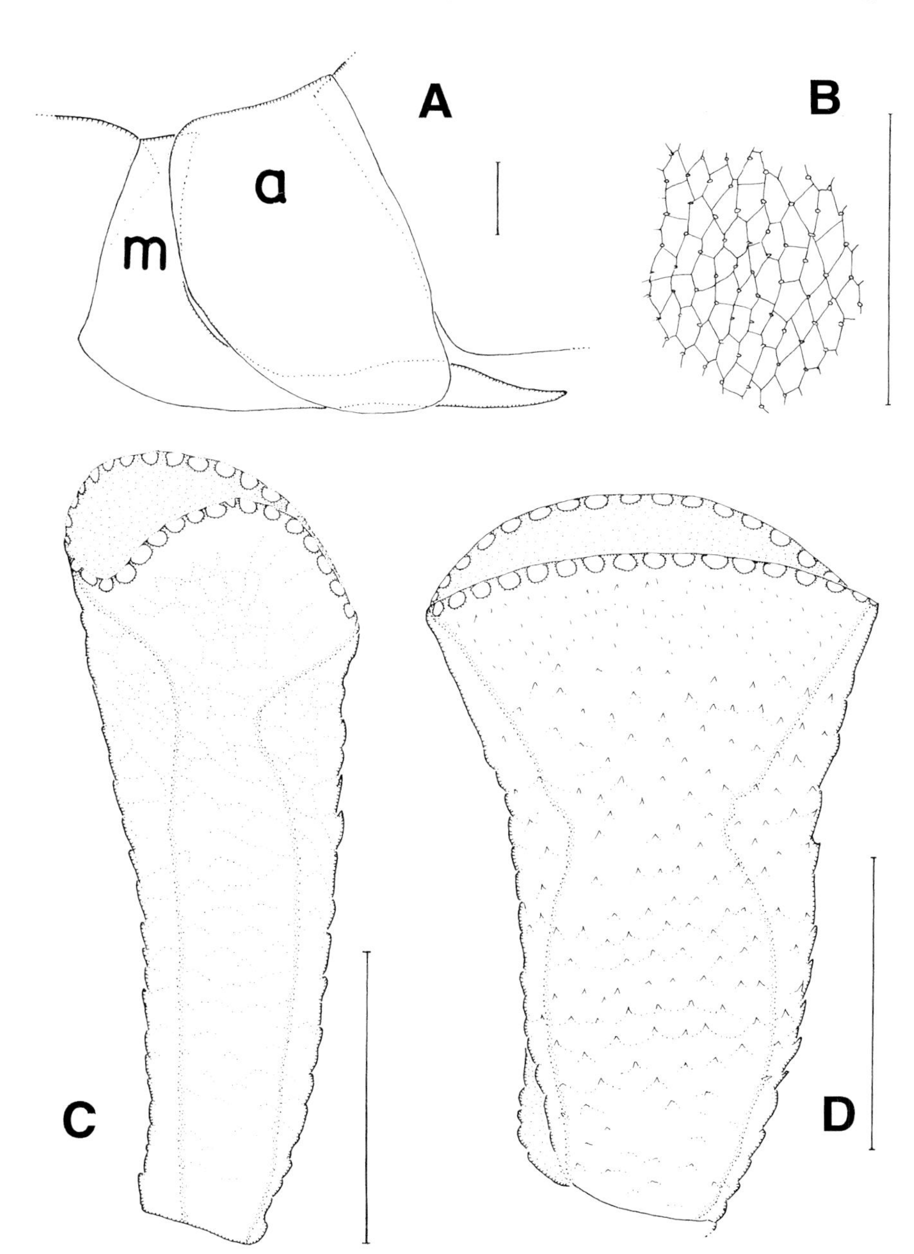
A
B
a
m
C
D

10 Respiratory trumpet less funnel shaped (Figs 17aes, obs; 16C)— **11**

— Respiratory trumpet more funnel shaped, being clearly widest at rim (Figs 17att, fil, mar, ser; 16D)— **13**

11 The basal section of the respiratory trumpet is long in relation to length of distal cup section (Fig. 16C)— **Dixella graeca** (Pandazis)

— The basal section of respiratory trumpet not so long in relation to cup section (Fig. 17aes, obs)— **12**

12 No trace of network of ridges linking denticles on sides of upper half of first abdominal segment. Respiratory trumpet as in Fig. 17obs— **Dixella obscura** (Loew)

— Usually a network of ridges (similar to Fig. 16B) linking denticles on upper half of first abdominal segment, but critical lighting may be required in order to see them. Respiratory trumpet as in Fig. 17aes— **Dixella aestivalis** (Meigen)

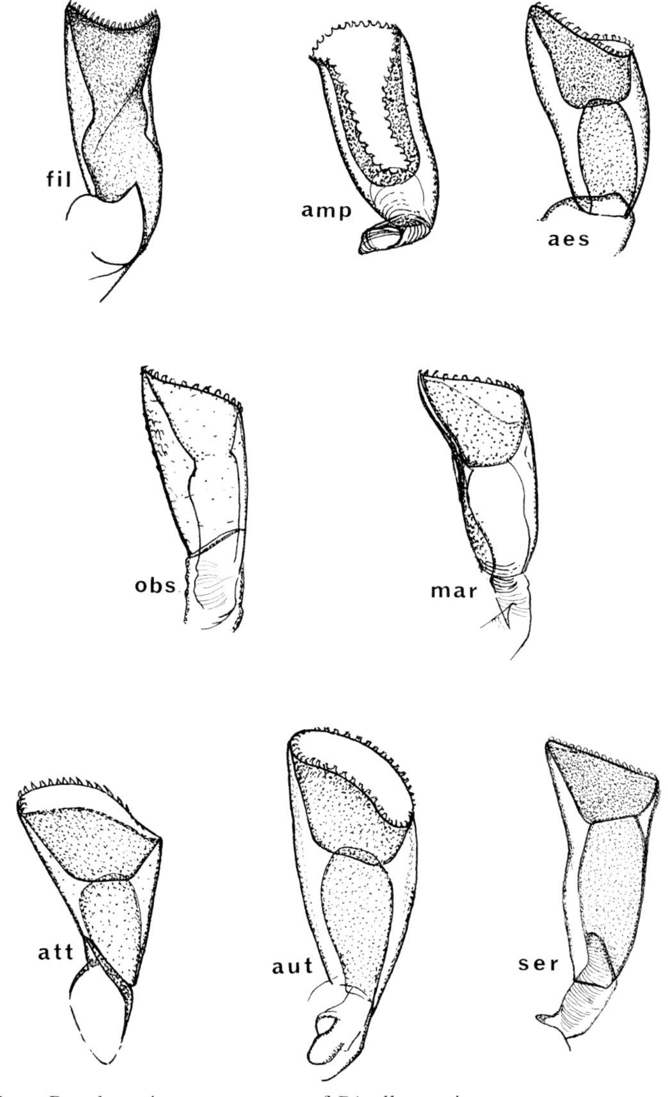

Fig. 17. Pupal respiratory trumpets of *Dixella* species.
Top row: fil, *D. filicornis*; amp, *D. amphibia*; aes, *D. aestivalis*.
Middle row: obs, *D. obscura*; mar, *D. martinii*.
Bottom row: att, *D. attica*; aut, *D. autumnalis*; ser, *D. serotina*.

13 The denticles on side of upper half of first abdominal segment (in position of the 'a' in Fig. 16A) are linked by a fine network of ridges (as in Fig. 16B). Respiratory trumpet as in Fig. 17att. Caudal lobes as in Fig. 18A— **Dixella attica** (Pandazis)

— These denticles not linked by a network of fine ridges, but disconnected ridges may be present. Respiratory trumpets and caudal lobes differ in detail (Figs 17fil, mar; 18B,C)— **14**

14 Caudal lobes with stronger teeth (Fig. 18B). Respiratory trumpets as in Fig. 17fil; 16D)— **Dixella filicornis** (Edwards)

— Caudal lobes with weaker teeth (Fig. 18C). Respiratory trumpet as in Fig. 17mar— **Dixella martinii** (Peus)

Fig. 18. (*On facing page*). Tips of left caudal lobes (from above) in *Dixella* pupae. **A**, *D. attica*; **B**, *D. filicornis*; **C**, *D. martinii*. (Scale bar = 0.1 mm).

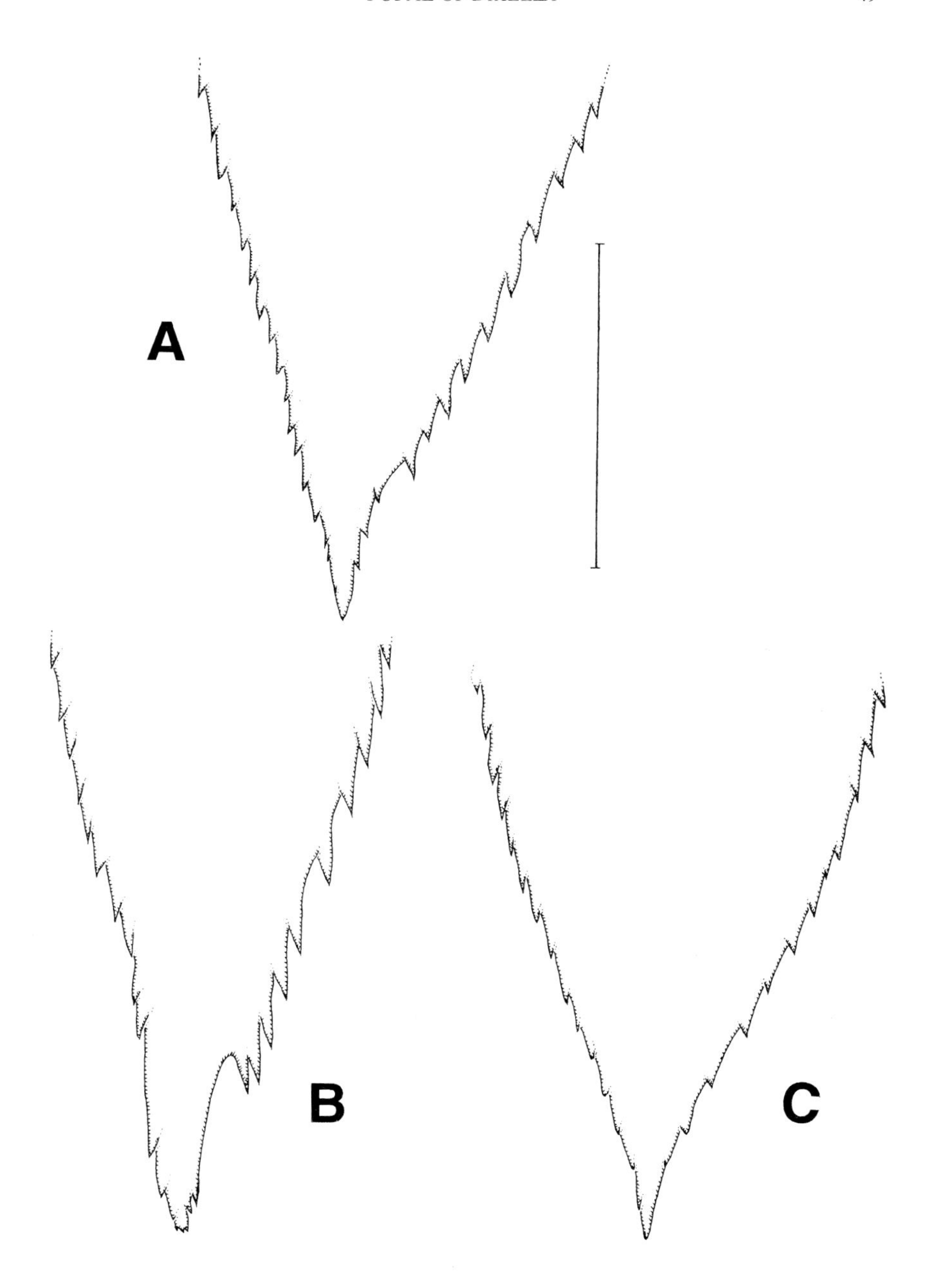
A
B
C

KEY TO ADULTS OF DIXIDAE

The abdominal terminalia are essential for the recognition of adult meniscus midges. While the following key mostly employs other characters to separate the species, the features of the terminalia are the most reliable characters for confirming the identity of a species.

Pigmentation and precise positions of wing veins may vary (e.g. Plate 5, M and N). However I have, for example, specimens in which the precise position of the rm cross vein (see Fig. 21) is different on the two wings. Some species vary more than others. *Dixella autumnalis* is the most variable and this variation may indicate that in reality there may be two or more sibling species standing under this name. In the keys, some allowance for intraspecific variation has been accommodated by keying a species both ways at certain couplets. Pigmentation is the least reliable character. The dark stripes of the thorax are occasionally devoid of pigment, appearing as stripes that are paler than the rest of the thorax. More commonly the whole thorax is much darker than usual, to give so-called melanic specimens.

Figs 19 and 20 respectively illustrate adult *Dixa* and *Dixella*; the *parts of the thorax* are labelled in Fig. 20. Important *wing veins* are labelled in Fig. 21C and 21D. The *male terminalia*, along with tergite 9, are rotated through 180 degrees at the time of emergence. The principal appendages are the pair of claspers, each consisting of a basal coxite and an apical style (Fig. 22). The coxite bears one or more processes from its inner margin. The *female terminalia* include a taxonomically useful sternite 9, situated between the lateral (descending) processes at the rear of segment 9. Thus, viewed from behind, in Fig. 29C the U-like sternite, together with these lateral processes, form the M-like structure of a transverse bridge.

1 Abdomen terminating in a pair of claspers, each consisting of a coxite and style (Figs 22–27)— MALES, **2** (p. 56)

— Abdomen terminating in simple cerci (Figs 28–31, 33)— FEMALES, **17** (p. 68)

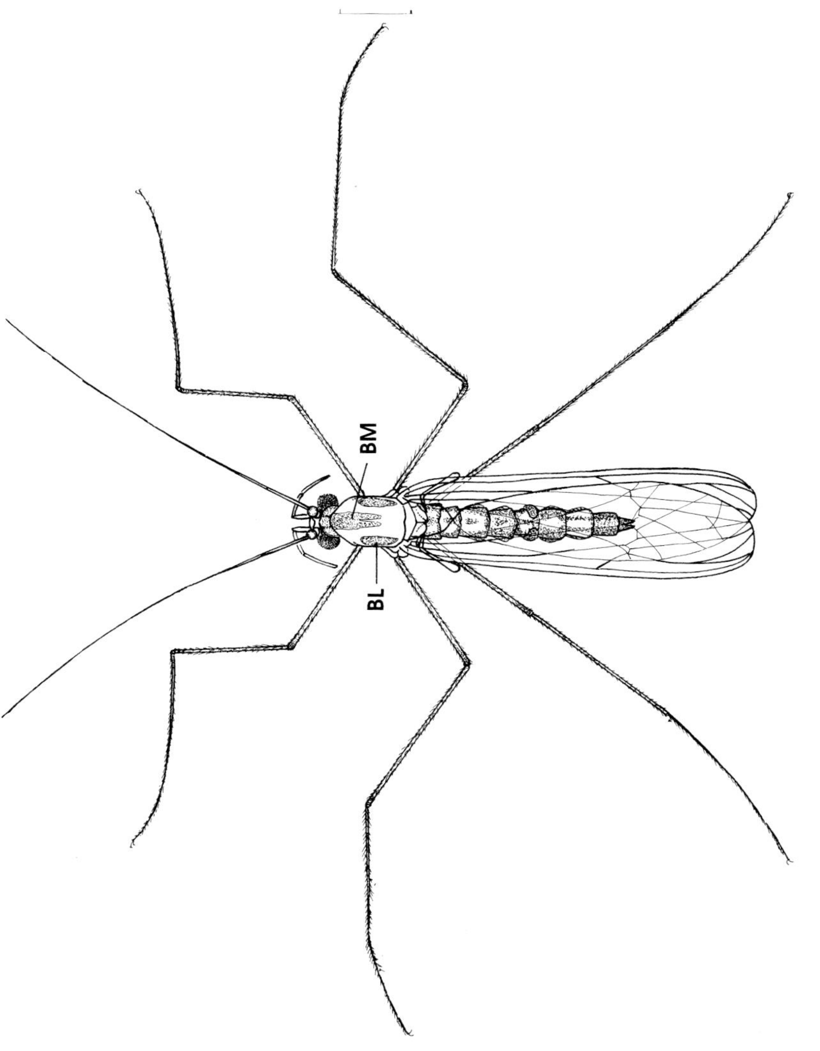

Fig. 19. Dorsal view of adult *Dixa*. BM, BL, median and lateral bands on the scutum.

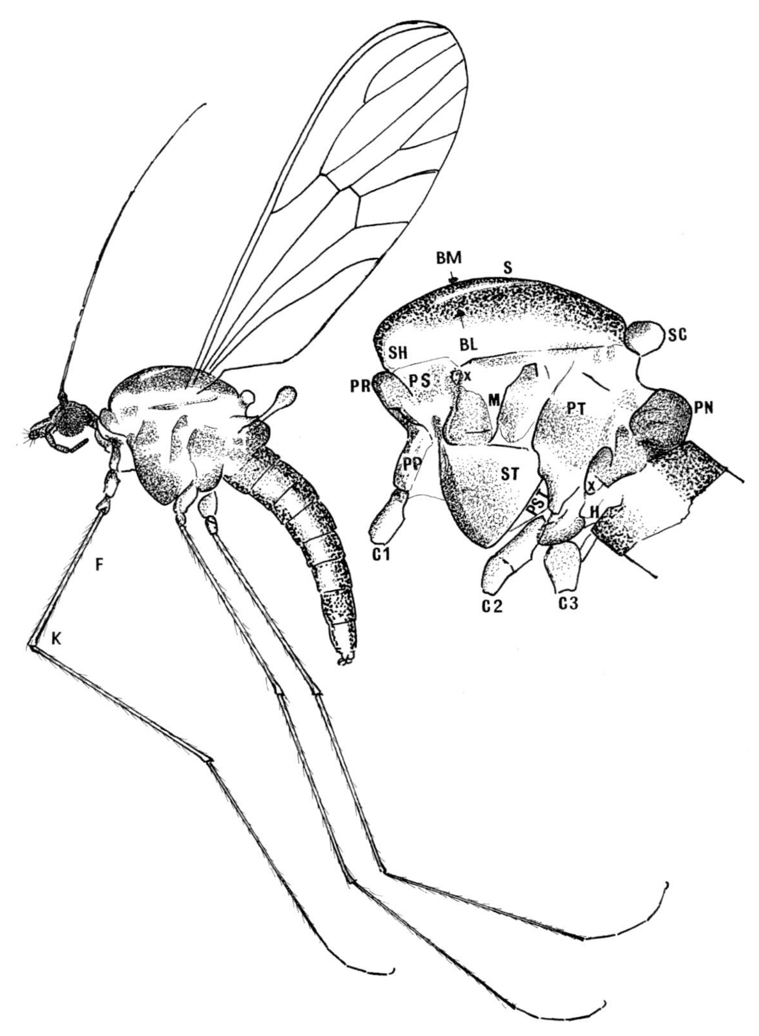

Fig. 20. Adult *Dixella* in side view. Left: whole insect, with appendages of one side only shown: F, femur; K, "knee". Right: enlargement of thoracic region. BM, BL, median and lateral bands on the scutum; C1–C3, coxae of legs; H, hypopleuron; M, mesopleuron; PN, postnotum; PP, propleuron; PR, pronotum; PS, prescutum; PST, post-sternopleural triangle; PT, pteropleuron; S, scutum; SC, scutellum; SH, shoulder of scutum; ST, sternopleuron; x, spiracle,

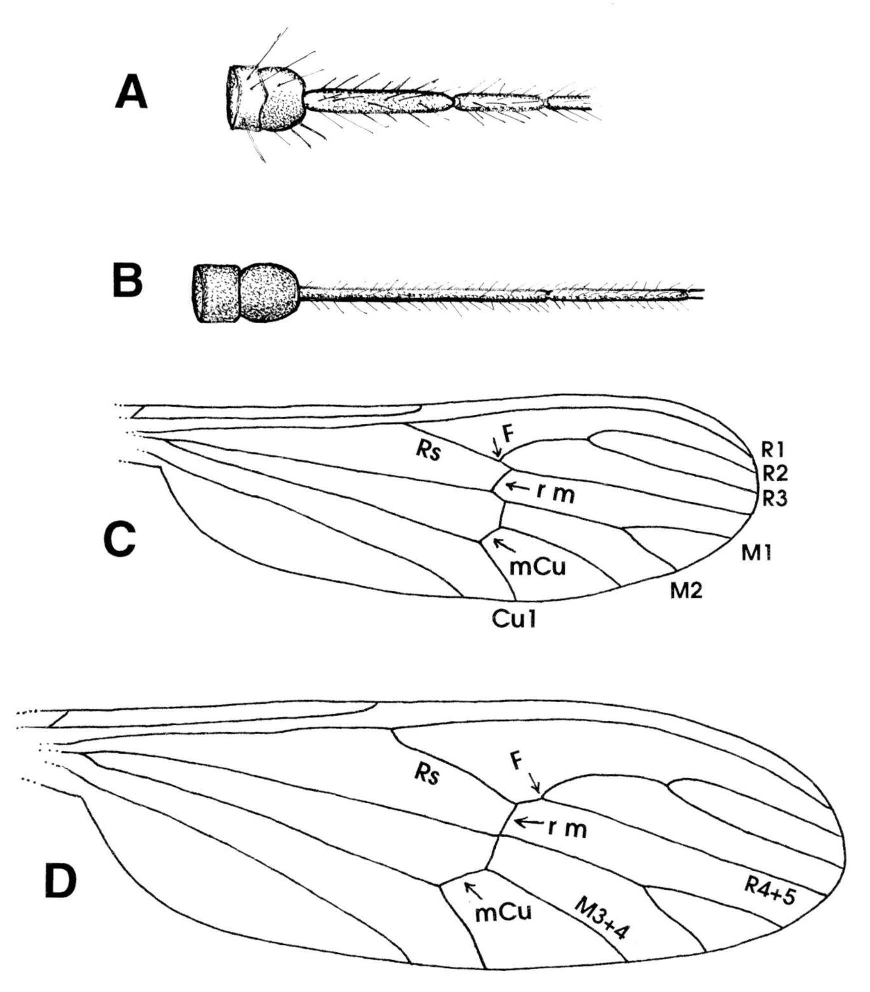

Fig. 21. **A, B** – basal segments of the antenna of: **A**, *Dixa;* **B**, *Dixella*. **C, D** – wings of Dixidae, with principal veins labelled: **C**, cross vein rm occurs *after* the fork (F) of vein Rs (i.e. towards the tip of the wing); **D**, cross vein rm occurs *before* the fork (F) of vein Rs (i.e. further towards the base of the wing).

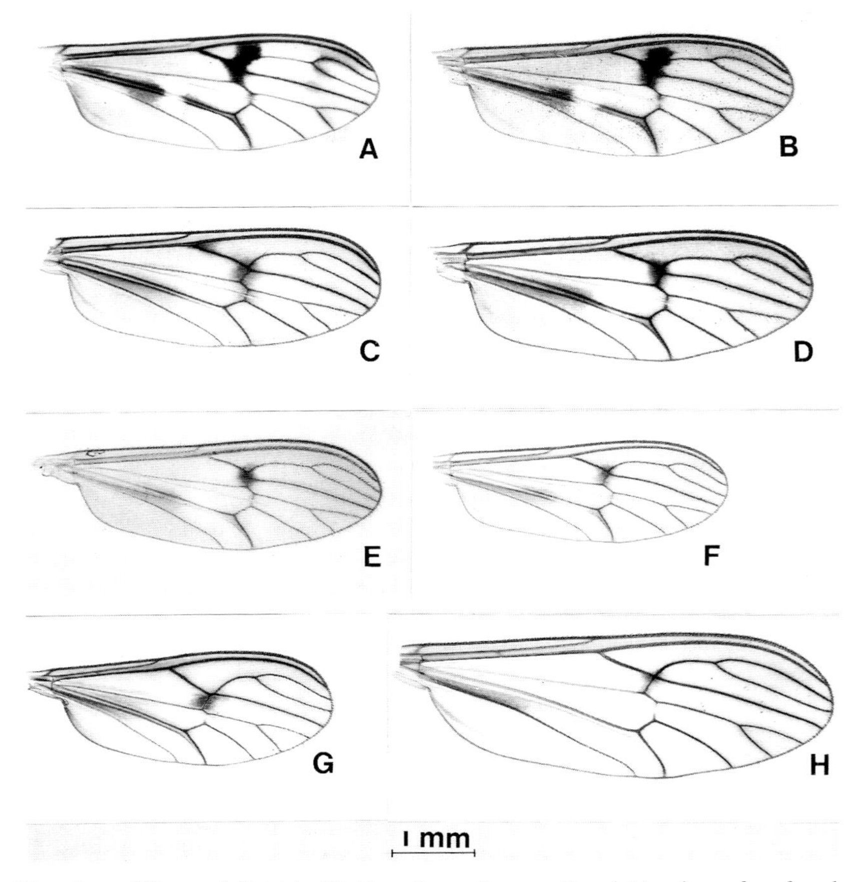

Plate 4. Wings of British Dixidae. In each case the right wing of a female is illustrated. **A**, *Dixa nebulosa*; **B**, *D. nubulipennis*; **C**, *D. puberula*; **D**, *D. submaculata*; **E**, *D. maculata*; **F**, *D. dilatata*, **G**, *Dixella filicornis*; **H**, *D. aestivalis*. (Photographs by A. E. Ramsbottom).

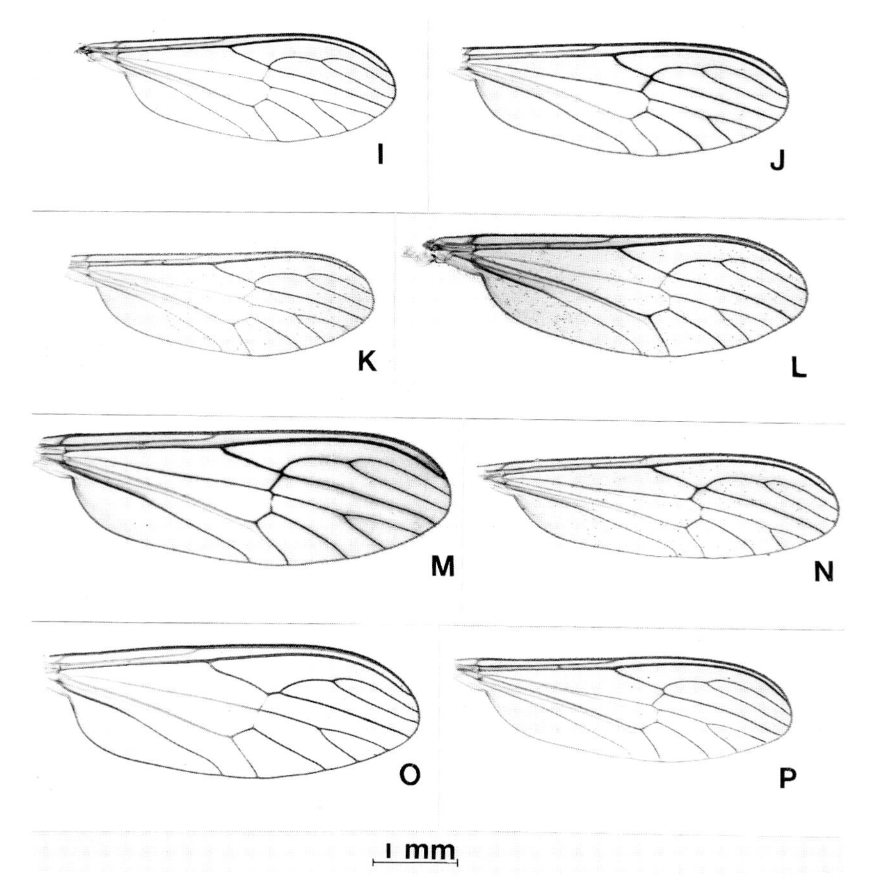

Plate 5. Wings of British Dixidae (*continued*). **I**, *Dixella martinii*; **J**, *D. serotina*; **K**, *D. attica* – atypical form; **L**, *D. attica* – usual form; **M**, *D. autumnalis* – usual form; **N**, *D. autumnalis* – atypical form; **O**, *D. obscura*; **P**, *D. amphibia*. (Photographs by A. E. Ramsbottom).

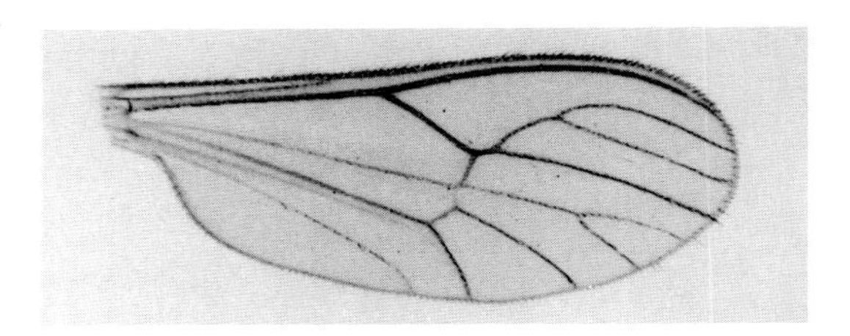

Plate 6. Wing of *Dixella graeca*. (Photograph by R. H. L. Disney).

KEY TO MALES OF DIXIDAE

2 (1, p. 50) Apical process of coxite small and inconspicuous (Figs 22↗↗p, 23C↗↗). Wings with dark clouds, at least over rm cross vein (Plate 4, A-F). Antennae relatively short, with relatively short and thick flagellar segments (Fig. 21A)— DIXA, **3**

— Apical process of coxite long, conspicuous and more-or-less opposed to style (Figs 24↗↗, 25↗↗, 26C, 27↗↗). Wings rarely with obvious dark clouds (Plate 4, G-H), more usually without or with very faint clouds (Plates 5 and 6). Antennae relatively long, with relatively long and slender flagellar segments (Fig. 21B)— DIXELLA, **8** (p. 60)

3 Dark cloud over cross vein rm large, well defined, and spreading forwards towards vein R1 (Plate 4, A and B). Sides of thorax distinctly striped horizontally (even in melanic forms this striping can be still discerned)— **4**

— This dark cloud is smaller, ill-defined, and not extending forwards to R1 (Plate 4, C–F). Sides of thorax not obviously striped— **5**

4 Forks of veins R2/R3 and M1/M2 clouded, though sometimes these clouds are faint (Plate 4, A). Without curved spines on apical corners of abdominal tergite 9 (Fig. 22A↗↗)— **Dixa nebulosa** Meigen

— These forks not clouded (Plate 4, B). A robust black curved spine at each apical corner of tergite 9 (Fig. 22B↗↗)— **Dixa nubilipennis** Curtis

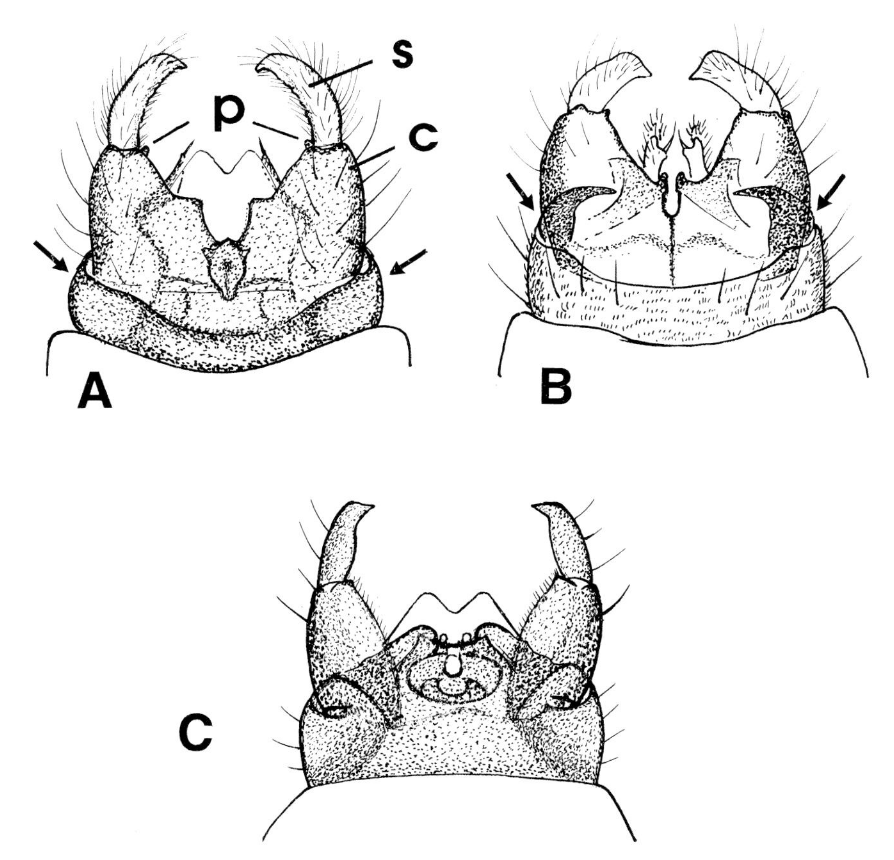

Fig. 22. Terminalia of *Dixa* males (viewed from above). **A**, *D. nebulosa* (s = style; c = coxite of the clasper; ↗p small apical process of coxite; ↗ curved spine absent). **B**, *D. nubulipennis* (↗ curved spine present); **C**, *D. puberula*.

5 Shoulder of thoracic scutum (SH in Fig. 20) with a distinct, but seemingly smudged, dark mark in front of lateral dark stripe of mesonotum. Typically with a small dark cloud over angle between veins R1 and Rs but no clouding behind the angle of Cu1 where this vein is joined by the cross vein mCu (Plate 4, C). Abdominal terminalia as in Fig. 22C— **Dixa puberula** Loew

— Shoulder of scutum without this dark mark. Typically without a trace of a cloud over angle between R1 and Rs but usually with some clouding in angle of Cu1 and mCu (Plate 4, D–F). Terminalia otherwise— **6**

6 With a prominent, black, two-branched median process on inner face of each coxite (Fig. 23B➚) and a black curved spine at each apical corner of abdominal tergite 9 (as in Fig. 22B➚)— **Dixa maculata** Meigen

— No such processes or spines— **7**

7 Longitudinal mesonotal bands separate (as in Fig. 19). Sides of thorax in part yellowish. Abdominal terminalia as in Fig. 23A— **Dixa submaculata** Edwards

— Anterior end of median band of mesonotum linked to front ends of lateral bands by ill-defined bands of dark pigment (as in Fig. 20). Sides of thorax entirely dark, even if paler in places. Terminalia as in Fig. 23C— **Dixa dilatata** Strobl

---

Fig. 23. (*On facing page*). Terminalia of *Dixa* males (viewed from above). **A**, *D. submaculata*; **B**, *D. maculata* (➚median process); **C**, *D. dilatata* (➚ small apical process of coxite).

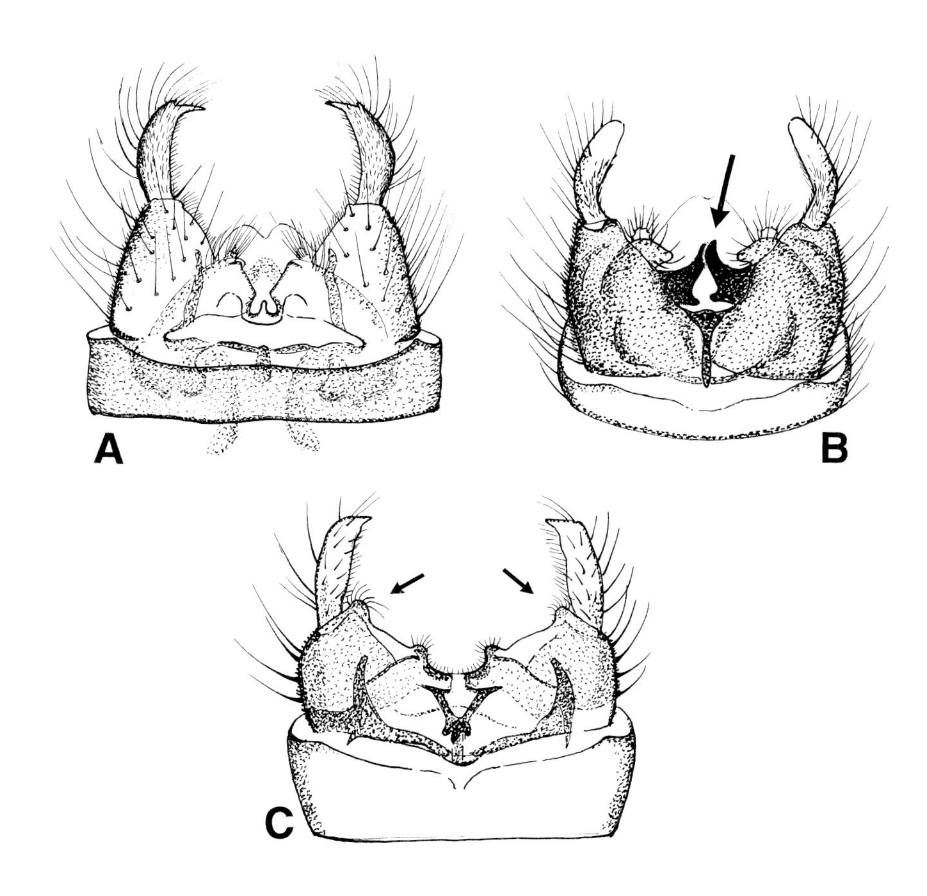
A
B
C

8 (2) Cross vein rm before the fork of Rs (Fig. 21D; Plates 4G, 5N,O and 6)— **9**

— Cross vein rm at or beyond the fork of Rs (Fig. 21C; Plates 4H, 5I–M and P)— **12**

9 Cross vein rm typically clouded (Plate 4G). Median dark band on top of thorax quite distinct from lateral bands (as in Fig. 19). Sternopleuron (ST in Fig. 20) almost entirely dark. Abdominal terminalia as in Fig. 24A— **Dixella filicornis** (Edwards)

— Without this combination— **10**

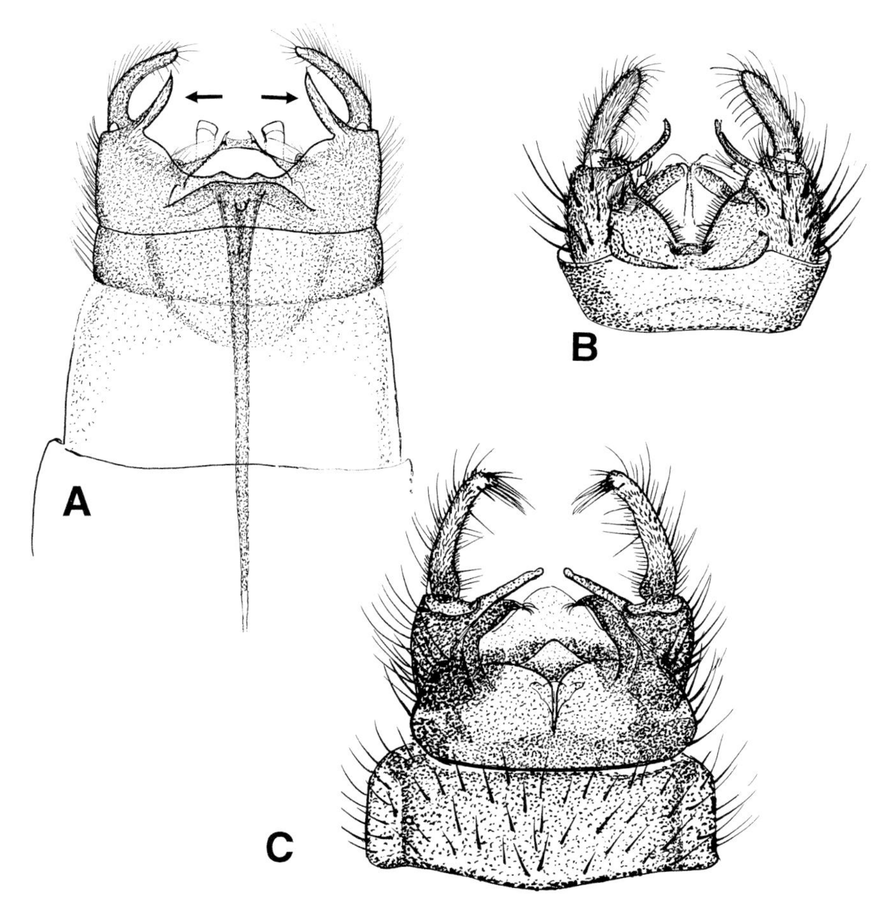

Fig. 24. Terminalia of *Dixella* males (viewed from above). **A**, *D. filicornis* (↗ apical process of coxite); **B**, *D. obscura*; **C**, *D. amphibia*.

10 Outer, postero-distal corner of front coxa with a pair of bristles (Fig. 25A↗). Abdominal terminalia as in Fig. 24B— **Dixella obscura** (Loew)

— No such bristles on front coxa (Fig. 25C). Terminalia otherwise— **11**

11 Frontal region (top) of head brown. Terminalia as in Fig. 25B— **Dixella graeca** (Pandazis)

— Frontal region of head orange to yellowish. Terminalia otherwise— **12**

12 (8 or 11) Thorax more-or-less dark all over. Styles long, relative to apical processes of coxites, and with a terminal tuft of hairs that confer a hooked appearance on each style (Fig. 24C)— **Dixella amphibia** (De Geer)

— Much of sides of thorax and at least the edges of mesonotom yellowish. Terminalia otherwise— **13**

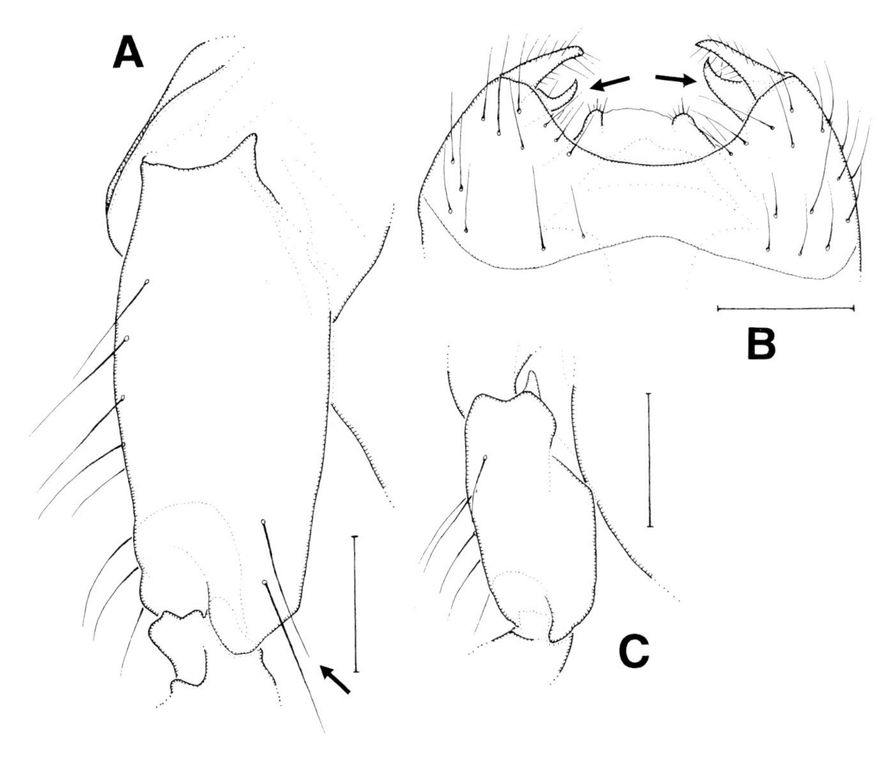

Fig. 25. Parts of *Dixella* males: **A**, left front coxa of *D. obscura* (↗ pair of bristles); **B**, terminalia of *D. graeca* (↗ apical process on style); **C**, left front coxa of *D. graeca*. (Scale bars = 0.1 mm).

13 The three longitudinal dark bands of mesonotum entirely separated from each other (as in Fig. 19). Apical processes of coxites forked (Fig. 26C↗fc)— **Dixella aestivalis** (Meigen)

— Front ends of lateral bands of mesonotum connected to anterior end of median band by less well-defined bands of dark pigment (as in Fig. 20). Apical processes of coxites unbranched— **14**

14 Outer margins of styles (see Fig. 22A) angled in distal halves (Figs 26A↗↗, B↗↗)— **15**

— Styles without this feature (Fig. 27) **16**

15 Abdominal terminalia as in Fig. 26A. The basal process of each coxite in the form of a dark, heavily sclerotised, curved spike (Fig. 26A↗sp). The bridge between the coxites is also heavily sclerotised— **Dixella serotina** (Meigen)

— Terminalia as in Fig. 26B. The basal process of each coxite smaller and paler. The bridge between the coxites is not heavily sclerotised— **Dixella autumnalis** (Meigen)

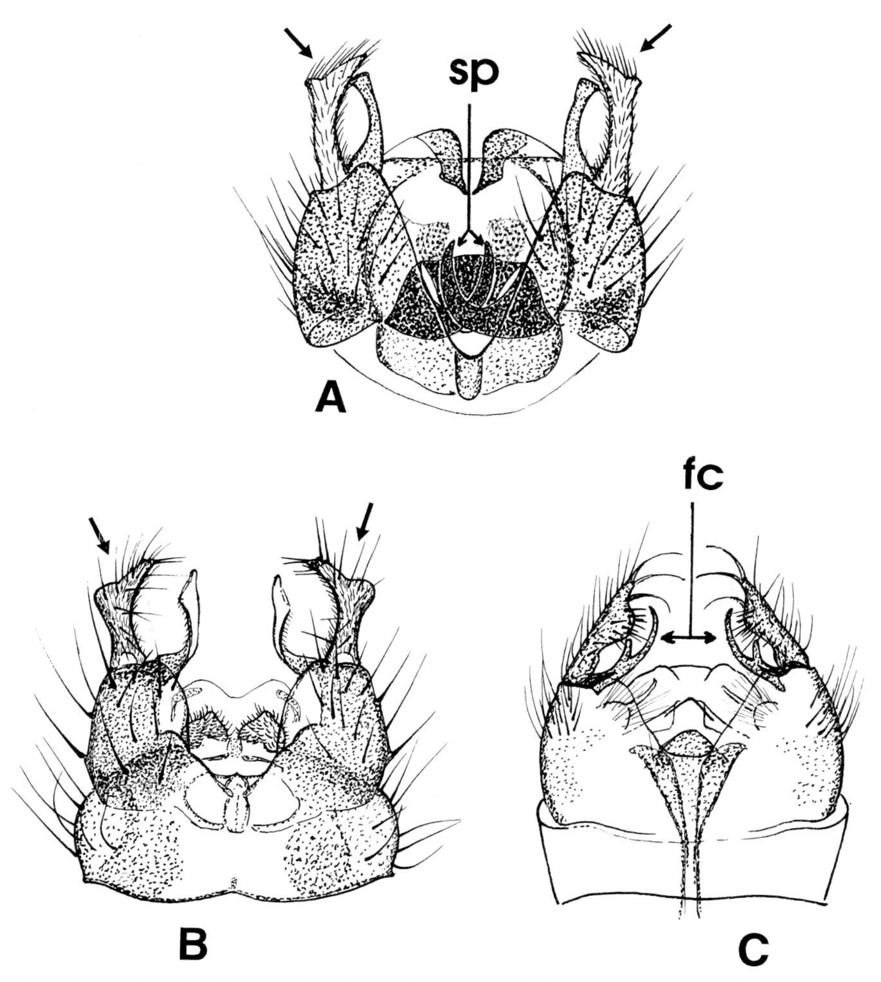

Fig. 26. Terminalia of *Dixella* males (viewed from above). **A**, *D. serotina* (↗ angle in style; ↗ sp curved spike); **B**, *D. autumnalis* (↗ angle in style); **C**, *D. aestivalis* (↗ fc forked apical processes of coxites).

16 Styles more finely tapered and apical process of coxite longer and more sinuous (Fig. 27A↗↗). Top of head dark brown to black—
Dixella martinii (Peus)

— Styles less finely tapered and apical process of coxite simply curved at base only (Fig. 27B↗↗). Top of head orange to dark brown—
Dixella attica (Pandazis)

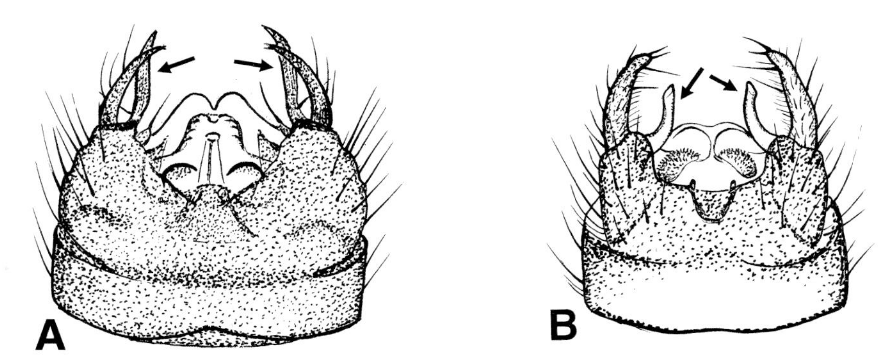

Fig. 27. Terminalia of *Dixella* males (viewed from above). **A**, *D. martinii* (↗ sinuous apical process of coxite); **B**, *D. attica* (↗ curved apical process of coxite).

KEY TO FEMALES OF DIXIDAE

17 (1, p. 50) Wing with at least one dark cloud, which is over the rm cross vein (Plate 4A–H). Even when faint, this cloud is visible if the wing is viewed against a white background— **18**

— With no dark cloud over rm cross vein (Plate 5I–P, Plate 6)— **25**

18 Dark cloud over cross vein rm large, well-defined, and spreading forwards towards vein R1 (Plate 4A, B). Sides of thorax distinctly striped horizontally (even in melanic forms this striping can still be discerned)— **19**

This dark cloud is smaller, ill-defined, and not extending forwards to R1 (Plate 4C–F). Sides of thorax not obviously striped— **20**

19 Forks of veins R2/R3 and M1/M2 clouded, though sometimes these clouds are faint (Plate 4A). Abdominal sternite 9 a simple bow (Fig. 28A➚)— **Dixa nebulosa** Meigen

— These forks not clouded (Plate 4B). Sternite 9 with a complex star-shape in middle (Fig. 28C➚)— **Dixa nubilipennis** Curtis

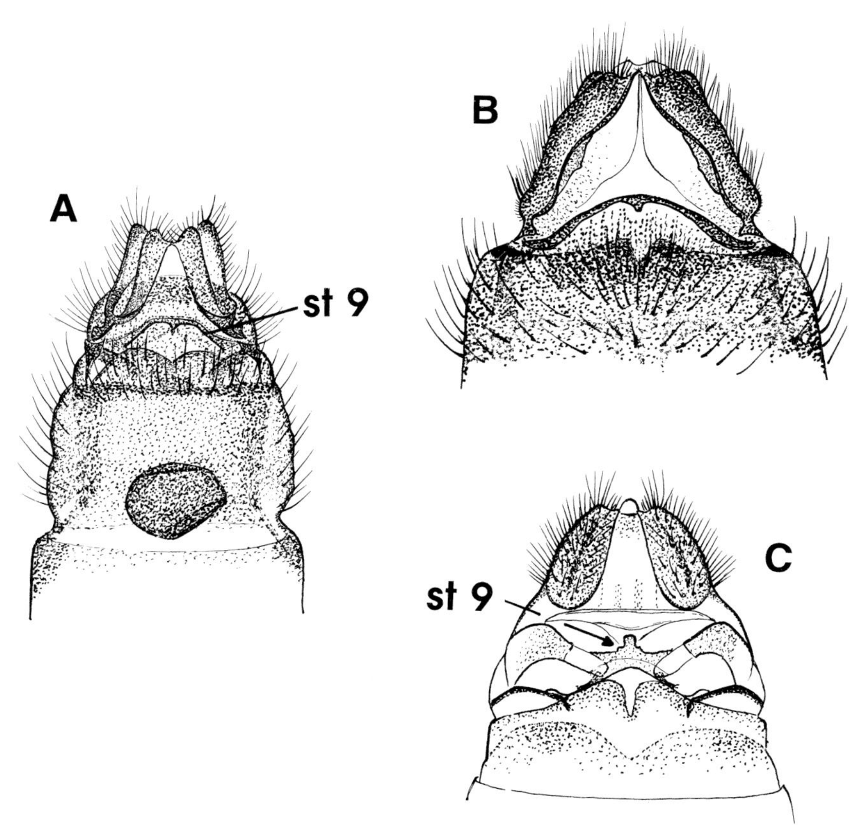

Fig. 28. Terminalia of *Dixa* females (ventral view). **A**, *D. nebulosa* (↗ simple bow of sternite 9; **B**, *D. puberula*; **C**, *D. nubulipennis* (↗ star-shape of sternite 9).

20 Cross vein rm at or before fork of vein Rs (Plate 4G). Median dark band of mesonotum quite separate from two lateral bands (as in Fig. 19). Sternopleuron (ST in Fig. 20) almost completely dark. Abdominal sternite 9 as in Fig. 30B— **Dixella filicornis** (Edwards)

— Without the above combination of characters— **21**

21 Antennae relatively short, with relatively short and thick flagellar segments (Fig. 21A). Typically the tips of femora with darker bands so that the "knees" (K in Fig. 20) appear darker than adjacent parts of legs— DIXA, **22**

— Antennae relatively long, with relatively long and slender flagellar segments (Fig. 21B). Typically "knees" not obviously darker than adjacent parts of legs— DIXELLA, **25**

22 Shoulder of thoracic scutum (SH in Fig. 20) with a distinct, but seemingly smudged, dark mark in front of lateral dark stripe of mesonotum. Typically with a small dark cloud over angle between veins R1 and Rs but no clouding behind the angle of Cu1 where this vein is joined by the cross vein mCu (Plate 4C). Abdominal sternite 9 as in Fig. 28B— **Dixa puberula** Loew

— Shoulder of scutum without this dark mark. Typically without a trace of a cloud over angle between R1 and Rs but usually with some clouding in angle of Cu1 (Plate 4D–F). Sternite 9 otherwise (Fig. 29)— **23**

23 Anterior end of median band of mesonotum linked to front ends of lateral bands by ill-defined bands of dark pigment (as in Fig. 20). Sides of thorax entirely dark, even if paler in places. Sternite 9 as in Fig. 29C↗— **Dixa dilatata** Strobl

— Without these features combined— **24**

24 Abdominal sternite 9 as in Fig. 29A↗. Longitudinal mesonotal bands typically separate (as in Fig. 19). Sides of thorax in part yellowish— **Dixa submaculata** Edwards

— Sternite 9 as in Fig. 29B↗. Mesonotal bands typically separate but sometimes faint bands link their front ends. Sides of thorax with any yellowish parts restricted to upper halves— **Dixa maculata** Meigen

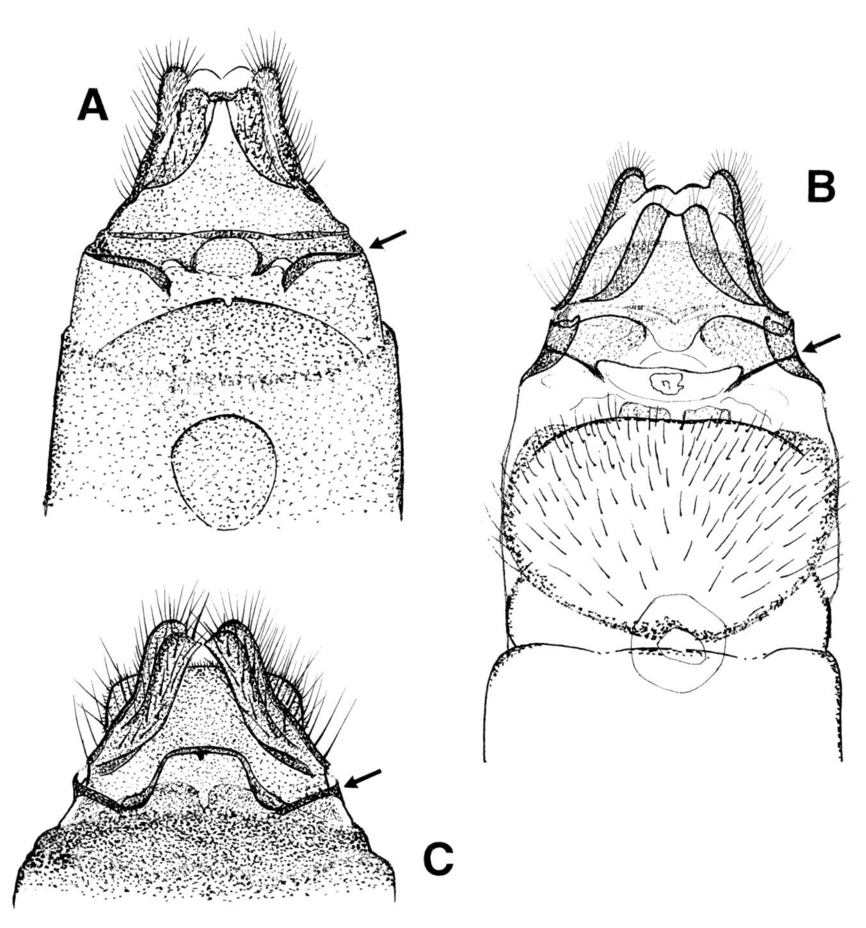

Fig. 29. Terminalia of *Dixa* females (viewed from above). **A**, *D. submaculata*; **B**, *D. maculata*; **C**, *D. dilatata*. (↗ sternite 9).

25 (17 or 21) With two or more bristles on outer postero-apical corner of front coxa (e.g. Fig. 25A➚)— **26**

— Without bristles in this position (e.g. Fig. 25C)— **28**

26 The three longitudinal dark bands of mesonotum quite separate from each other (as in Fig. 19). Abdominal sternite 9 as in Fig. 30B➚—
Dixella filicornis (Edwards) var. **immaculosa** Sicart

— Median dark band of mesonotum linked to two lateral bands by ill-defined bands from front ends of the latter (as in Fig. 20). Sternite 9 otherwise— **27**

27 Internally the atrium with a pair of dark brown to black sclerotised hinge teeth between the dark spherical spermatheca and sternite 9 (Fig. 31B➚ ht)— **Dixella attica** (Pandazis)

— Without sclerotised hinge teeth in atrium, and sternite 9 as in Fig. 32A, but sometimes obscure in middle (Fig. 30A➚)—
Dixella obscura (Loew)

28 (25) Thorax darkened all over. Internally, in the atrium behind the dark spherical spermatheca, there is a distinctive sclerotised insula and behind this two short, parallel rows of spinules (Fig. 30C➚)—
Dixella amphibia (De Geer)

— Much of sides of thorax yellowish. Without insula and two rows of spinules in atrium— **29**

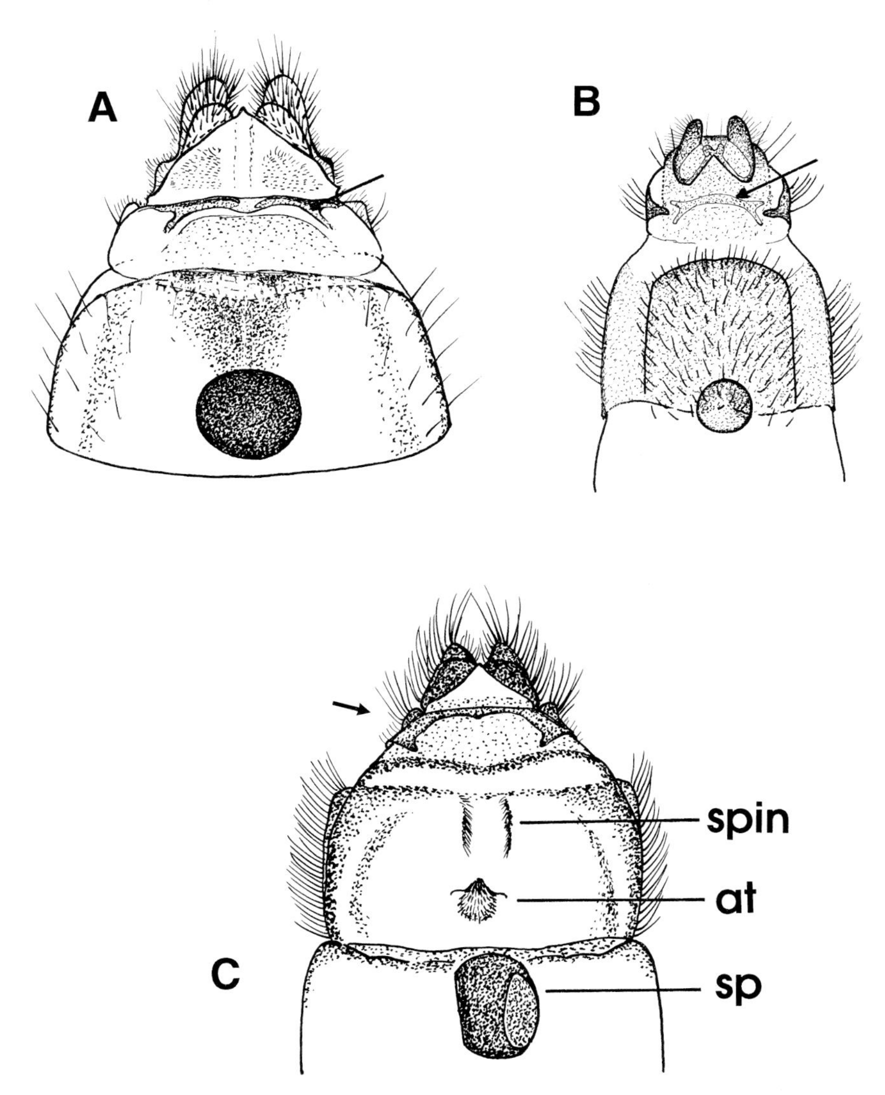

Fig. 30. Terminalia of *Dixella* females (ventral view). **A**, *D. obscura* (↗ sternite 9); **B**, *D. filicornis* (↗ sternite 9); **C**, *D. amphibia* (↗ at atrium; ↗ sp spermatheca; ↗ spin spinules; ↗ sternite 9).

29 Median and lateral longitudal dark bands of mesonotum quite separate from each other (as in Fig. 19). Internally the atrium has an irregular patch of small spinules, behind the dark, spherical spermatheca and between a pair of weakly-developed hinge teeth (Fig. 31C↗). Typically with an obscure cloud over cross vein rm (Plate 4H)—

Dixella aestivalis (Meigen)

— Front ends of lateral bands of mesonotum linked to anterior end of median band by ill-defined bands of dark pigment (as in Fig. 20). No cloud over cross vein rm— **30**

30 Hinge teeth conspicuous in atrium (Fig. 31A↗). Top of head black—

Dixella martinii (Peus)

— Hinge teeth not sclerotised. Top of head frequently, in part at least, yellowish to orange, but may be brown— **31**

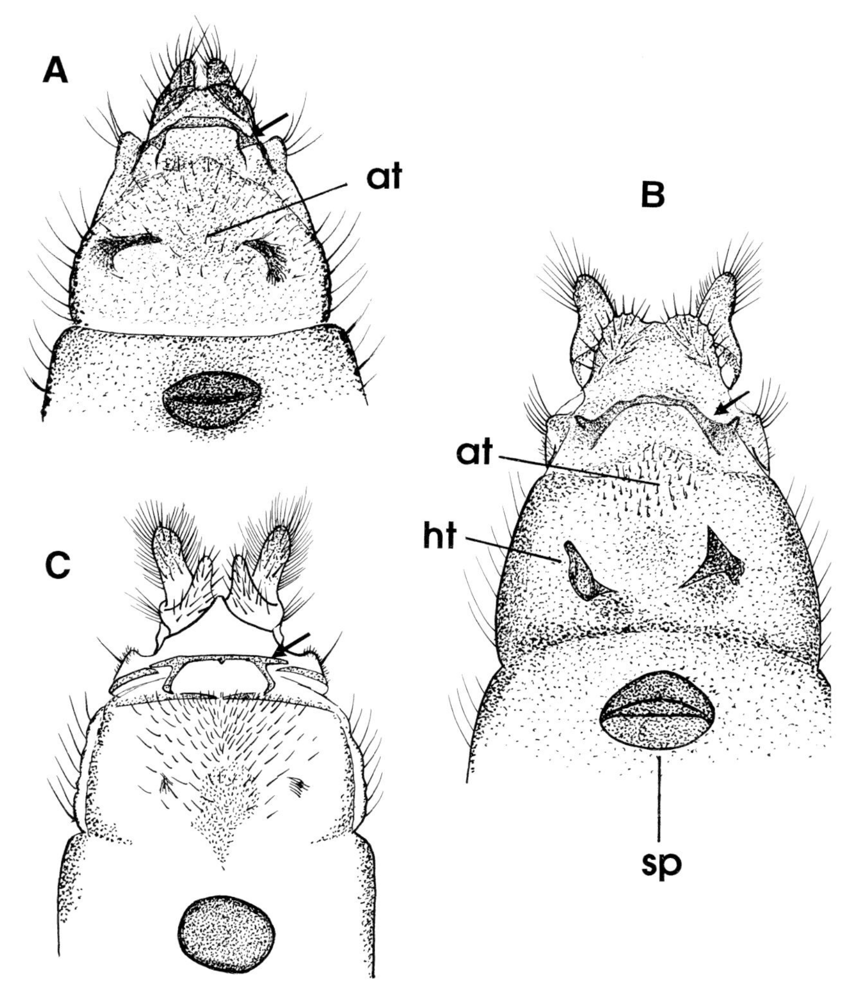

Fig. 31. Terminalia of *Dixella* females (ventral view). **A**, *D. martinii* (↗ sternite 9; ↗ at atrium); **B**, *D. attica* (↗ sternite 9; ↗ at atrium; ↗ ht hinge teeth; ↗ sp spermatheca); **C**, *D. aestivalis* (↗ sternite 9).

31 Top of head brown. Sides of thorax largely dark. Abdominal sternite 9 as in Fig. 32B— **Dixella graeca** (Pandazis)

— Top of head, at least along median band of frons, yellowish or orange. Sides of thorax in part yellowish. Sternite 9 otherwise (Fig. 33)— **32**

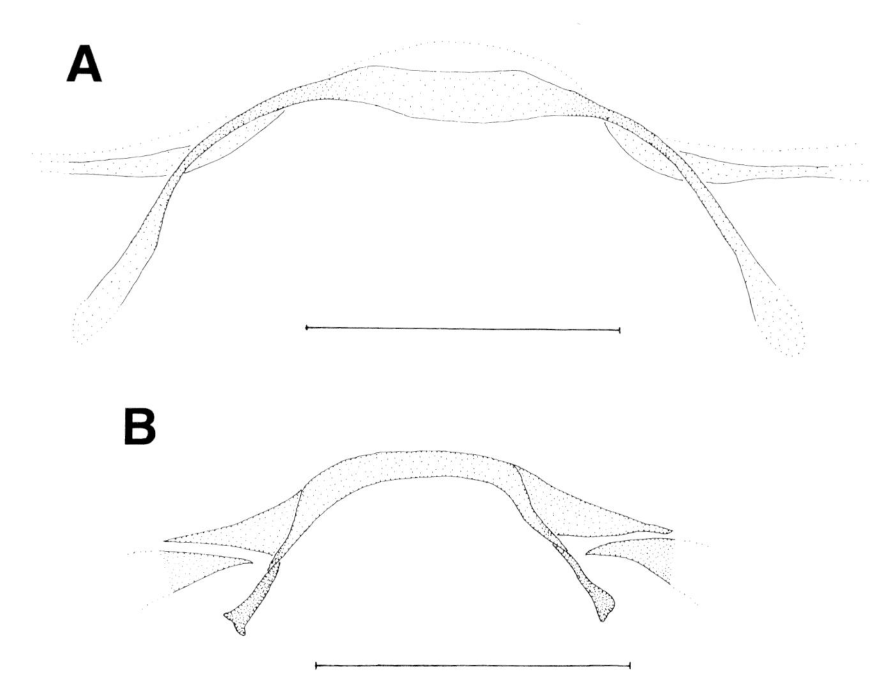

Fig. 32. Abdominal sternite 9 of *Dixella* females. **A**, *D. obscura*; **B**, *D. graeca*. (Scale bars = 0.1 mm).

32 With a distinctive sternite 9 (Fig. 33A) and rest of abdominal terminalia as in Fig. 33A, but the postero-lateral clusters of spinules on segment 8 may be reduced— **Dixella autumnalis** (Meigen)

— With a distinctive sternite 9 (Fig. 33 B↗) and rest of terminalia as in Fig. 33B— **Dixella serotina** (Meigen)

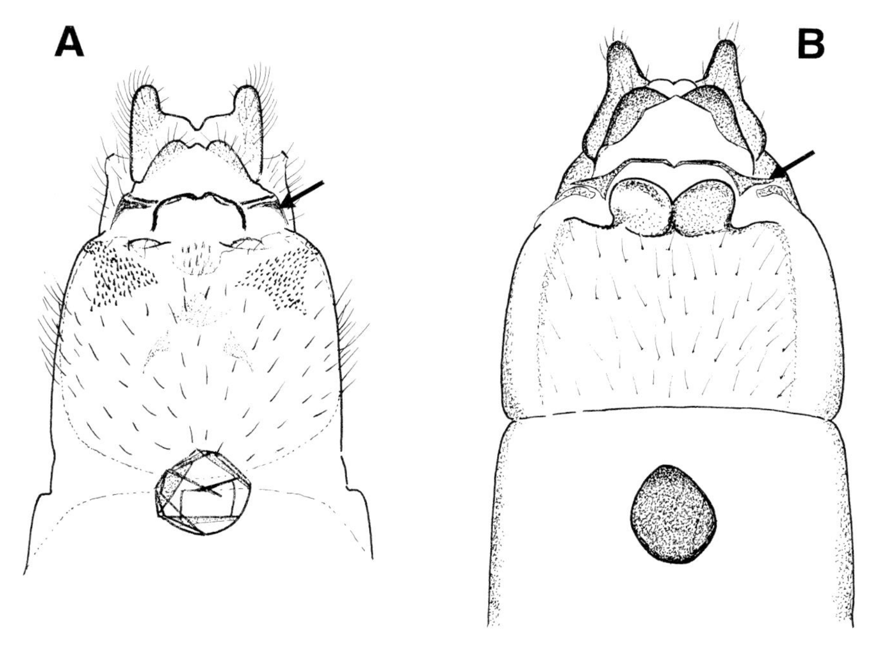

Fig. 33. Terminalia of *Dixella* females (ventral view). **A**, *D. autumnalis* (↗ sternite 9); **B**, *D. serotina* (↗ sternite 9).

ECOLOGY AND NOTES ON SPECIES OF DIXIDAE

Larvae feed on micro-organisms and particles of decaying organic matter suspended in the water or trapped in the surface film. Miall (1895) provides a good account of the mode of feeding of a *Dixa* larva. This account should be supplemented by referring to Guthrie (1989), who provides an excellent introduction to the physical properties of the meniscus. She then proceeds to emphasise that on natural bodies of still water the interface between water and air tends to be a "skin" that behaves like a monomolecular layer of protein and lipoprotein. Furthermore, this surface film tends to accumulate many other organic molecules, bacteria, protozoa and other micro-organisms. The result is a highly nutritious resource for insects able to exploit it. While all meniscus midge larvae feed on microscopic particles and organisms suspended in the water, in a manner similar to mosquito larvae, *Dixella* larvae in particular also ingest the surface film (see Fig. 24 in Guthrie 1989).

Overwintering larvae are able to withstand severe cold. For example Wright (1901) reported healthy *Dixella* larvae in pools in Scotland that had been iced over and covered with 45–60 cm of snow for two weeks. At the time of the investigation the lowest layer of the snow-cover was frozen hard.

Mortality factors for larvae have not been systematically studied. In laboratory cultures a cytoplasmic polyhedrose virus (Plate 7) can be devastating (Goldie-Smith 1987). General predators, such as dytiscid beetle larvae and damselfly nymphs (e.g. Nicholson 1978), take their toll, but detailed studies of selectivity and the impact on dixid larval populations remain to be done.

Dixa larvae are sometimes a significant component of the invertebrates occurring in the downstream drift in streams and rivers, with peak numbers at night (e.g. Elliott & Minshall 1968; Elliott & Tullett 1977). This nocturnal peak is possibly due to the mature larvae seeking suitable sites for pupation chiefly at night. However, increased disturbance by night-active elements of the associated stream fauna would also explain the enhanced night-time drift of *Dixella* larvae, as has been suggested by limited experiments with *Simulium* larvae (Disney 1972). Elliott (1967) gives a detailed account of the methods used to study downstream drift. Wagner (1980) reported on the numbers of adult Dixidae obtained in an emergence trap set in a stream on the Continent.

Thomas (1979) showed that *Dixa puberula* is an extremely sensitive indicator of the presence of surfactant or oil-borne pollutants in streams. Fowler, Withers & Dewhurst (1997) report that larval mortality in three species (*Dixa nebulosa, Dixella aestivalis* and *D. serotina*) correlates with decreasing surface tension caused by surfactant pollutants, and they conclude that dixid larvae are good indicators of such pollutants.

Preliminary observations on the timing of the emergence of adults from their

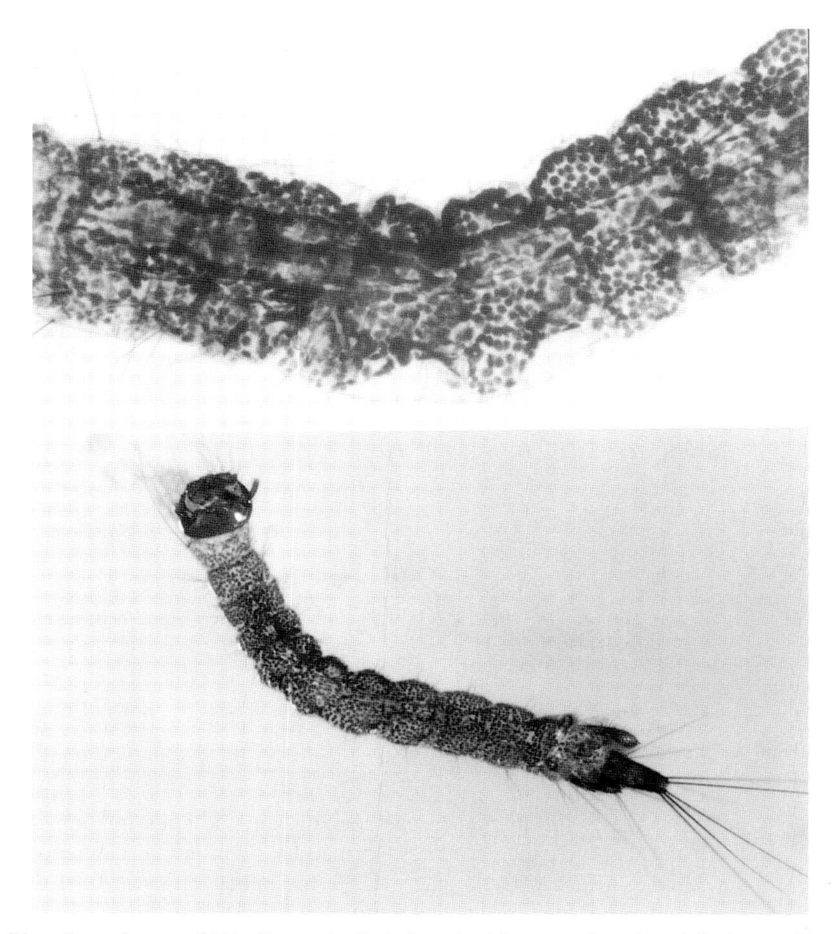

Plate 7. Larva of *Dixella aestivalis* infected with a cytoplasmic polyhedrose virus, showing the characteristic measled appearance of a whole larva (below) and enlarged part (above). (Photographs by E. K. Goldie-Smith).

pupae have reported that *Dixella aestivalis* emerged at night, after midnight, but *D. filicornis* emerged in the early morning (Morgan & Waddell 1961).

Reared adults of *Dixella* will readily mate in laboratory cultures, but *Dixa* fail to do so and probably require the males to form an aerial swarm (e.g. Goldie-Smith 1990a). Such swarms of *Dixa* males are probably most frequent around dusk and typically a female entering the swarm can be observed dropping out once she is coupled with a male; copulation is then completed on the ground. However, males of *Dixa puberula* have often been observed in loose swarms during the day, although Elliott & Tullett (1977) reported that male swarms of this species were mainly at dusk over or near water and that mating took place in flight. An aerial swarm of *Dixella autumnalis* males has also been recorded on a sunny day in winter (January 24th) (Goldie-Smith 1989a). However, while most *Dixella* species probably mate during the day, most will do so without the males first needing to swarm. *Dixa* adults feed on plant sugars in the laboratory (Goldie-Smith 1989a).

Adult Dixidae may have larval water-mites attached to them, usually on the abdomen. These mites are typically red and belong to the genus *Arrenurus* (Jones 1965). Stechmann (1977, 1978) has laid the foundations for studies of host specificity.

Since publication of the provisional distribution maps in the first edition (Disney 1975), much additional information has been accumulated and most is summarised in the revised maps published by Goldie-Smith (1990b). Irish records are given by Ashe (1986) and Ashe & O'Connor (1990). In the notes below, I provide summary statements largely based upon these publications.

Dixa dilatata

Larvae recorded March–October; adults February–November. The recorded larval habitats include trickles, spring flushes and small streams with emergent stones, rushes, sedges (e.g. *Carex riparia*) or dead leaves and the margins of larger streams; typically in open, exposed landscapes. Also recorded from peat pools, a reed (*Phragmites*) swamp, alder carr, a woodland pond and a muddy pool in a cave. Water chemistry of habitats ranges from acid peat-stained and iron-stained waters to soligenous flushes and a recorded pH as high as 7.1.

Distribution. Scattered throughout mainland Britain and Ireland, being commoner in the wetter parts. Also recorded from the Isle of Skye.

Dixa maculata

Larvae recorded May, August–October; adults January–March, May–October. The recorded larval habitats are stony woodland streams and a canal at the point of discharge of a piped stream.

Distribution. A generally rare species that is mainly restricted to scattered localities in the northern and western halves of England (with one record from Suffolk), and Wales.

Dixa nebulosa

Larvae and adults recorded March–December. The recorded larval habitats range from stony woodland streams to reed-beds at the margins of rivers, in the open. The larvae are typically found in emergent and trailing beds of grasses and rushes (especially *Agrostis stolonifera* and *Juncus effusus*, according to Fowler 1984a), fringing small lowland streams and dykes. They also occur on emergent stones at the edges of upland torrent streams, in canals and in watercress beds. Some records are from the fringes of still waters, such as a deep castle-moat, at the edge of a large lake, and from a fishpond dug in clay.

Peach & Kinsler (1988) give some data on seasonal changes in sex ratios, reporting a predominance of females in late summer emergences. Whether parthenogenesis is involved remains to be investigated. Peach (1984) reported adult eclosions mainly between 1400 and 0200 hours and adult activity was greatest around dusk. Wagner (1980) reports the numbers of adults obtained in an emergence trap on the Continent. The mortality of larvae due to surfactant pollutants is reported by Fowler, Withers & Dewhurst (1997).

Distribution. Scattered localities throughout mainland Britain and Ireland, and from the Isle of Man. *D. nebulosa* is the most frequently encountered species of *Dixa* in lowland England (Peach 1984).

Dixa nubilipennis

Larvae and adults recorded January–December, both with a peak in the Midlands in September–November. The recorded larval habitats are typically emergent dead leaves and stones in shallow, shaded streams, especially woodland. Also reported from small roadside and moorland streams, the inlet stream of a watercress bed, a calcareous spring, a spring-fed brackish ditch and in emergent vegetation in a canal at the point of discharge of a piped stream.

Roper (1962) recorded a male swarm in January and adults were attracted to light in September.

Distribution. Scattered localities throughout mainland Britain and Ireland. *D. nubilipennis* is the second most frequently encountered species of *Dixa* in lowland England (Peach 1984)

Dixa puberula

Larvae recorded January–December; adults March–December. Recorded larval habitats are typically stony torrent streams and small stony rivers both in woodland and in the open. In addition to their occurrence on emergent stones they can be found in leaf packets trapped against stones, amongst emergent patches of moss on boulders and in marginal patches of emergent vegetation. Occasionally found in emergent vegetation (e.g. sedges) immediately downstream of more stony reaches.

Elliott & Tullett (1977) provide quantitative data on fluctuations in the numbers of this species in the downstream drift in two stony streams. The numbers of larvae drifting are greater at night than during the day, with peak drift early in the night. They also report that male swarms are mainly at dusk, over or near water. Mating takes place in flight. Elliott & Tullett observed oviposition into organic material in a small pool at the side of a stream. Pupae were typically 5–10 cm above the water surface; pupation occurs mainly at night. The pupal period was typically 2–4 days. Adults emerge both in the day and at night. Thomas (1979) adds information on this species as a sensitive indicator of pollutants, especially surfactants and oil-borne substances.

Distribution. Throughout mainland Britain and Ireland, but rare to the east of a line from Dorset to the Wash.

Dixa submaculata

Larvae recorded January–December; adults January–October. Recorded larval habitats are typically on emergent stones and dead leaves in shallow streams both in woodland and in the open. Also recorded on sedges in slow-flowing woodland streams and in marginal beds of trailing grasses fringing other small streams. Recorded in a canal at the point of discharge of a piped stream. Water chemistry ranges from calcareous to base-poor moorland streams, and from iron-rich to brackish waters. Hydrological conditions range from small trickles and slow moving ditches to rapid, stony torrent streams.

Wagner (1980) reports the numbers of adults obtained in an emergence trap on the Continent.

Distribution. Throughout mainland Britain and Ireland, and Isle of Wight. Also recorded from the Channel Islands.

Dixella aestivalis

Larvae recorded April–November, with a peak June–September; adults April–December. Recorded larval habitats include a wide range of emergent vegetation (sedges, rushes – especially *Juncus effusus*, reeds, *Typha, Sparganium erectum, Equisetum, Glyceria fluitans, G. maxima, Potamogoeton* and *Elodea*) in a variety of pools, ponds, lake margins, swamp-filled hollows, woodland hydrosere swamps, river margins and backwaters, and dykes; in waters ranging from oligotrophic peat pools to alkaline canals. Especially characteristic of eutrophic waters.

Nicholson (1979) reported that at 19–21°C, larval instars 3 and 4 each lasted about 20–25 days and the pupal period was 2 days. Goldie-Smith (1989b) reported clutch sizes of 21 to 130 eggs; the first clutches laid were the largest. Large clutches were commoner in May and October. The mortality of larvae due to surfactant pollutants is reported by Fowler, Withers & Dewhurst (1997). Adults have been reported in emergence traps between midnight and 0400 hours in Scotland (Morgan & Waddell 1961). Stechmann (1977, 1978) recorded this species being parasitised by larvae of the mites *Arrenurus globator* and *A. truncatellus.*

Distribution. Throughout mainland Britain and Ireland, especially in the lowlands of England.

Dixella amphibia

Larvae recorded January–December; adults March–November. Recorded larval habitats include a range of sedge and reed swamps from woodland hydroseres and fens to acid bog pools and dykes, a brackish ditch, beds of iris, rushes, reeds, fallen leaves at the margins of ponds and lakes, and watercress beds. Associated plants include *Glyceria, Typha, Juncus, Molinia, Equisetum, Carex, Oenanthe, Myrica, Menyanthes* and *Potamogeton*.

Up to 90 eggs have been reported in a clutch, but typically there are 30–40 (Goldie-Smith 1989a). This species is parasitised by larvae of the mite *Arrenurus truncatellus* (Stechmann 1977).

Distribution. Throughout mainland Britain and Ireland, especially in the lowlands of England.

Dixella attica

Larvae recorded February–May, July–October; adults February–March, June–October. Larval habitats are mainly found near coasts and estuaries and include slightly brackish waters, such as dykes, with emergent sedges and grasses. Also recorded from marginal swamps bordering ponds, lakes and rivers. Associated plants include *Eleocharis, Apium, Scirpus maritimus, Ranunculus baudottii, Berula, Phragmites, Juncus, Glyceria fluitans* and *G. maxima*. Inland records are rare, but it has been collected from a rejuvenated old pond at Wembley (Williams & Fowler 1986) and a single adult was caught (by P. Holmes) about 40 km inland near Brecon; but Fowler (1989) endorses the primarily coastal and estuarine distribution of this species.

Distribution. Mainly near coasts and estuaries in mainland England, especially East Anglia and Kent, Wales and south-east Ireland. Also recorded from Bardsey Island (Fowler 1989).

Dixella autumnalis

Larvae recorded January–December, with a peak in March–May; adults recorded January–November. Larval habitats include swamps of grasses, reeds, rushes and sedges at the margins of lakes (especially those formed from old gravel pits, etc.), ponds, canals, ditches and rivers. Also recorded in watercress beds, dune slacks, fen swamps, more acid situations, and brackish ditches and ponds near the coast. Associated plants include *Phragmites, Carex, Juncus. Equisetum, Iris pseudacorus, Glyceria, Sparganium erectum* and *Elodea*. Adults have been recorded resting in a culvert pipe and in a light trap on the roof of a building.

Peach (1984) observed matings on vertical surfaces in late afternoon (when light levels were between 100 and 300 lux). Nicholson (1978) reported a pupal period of 2 days at 15°C. Peach & Fowler (1986) reported 106–119 eggs per clutch, these being reported above the water. Their incubation was 5 days at room temperatures (8 to 19°C) but at 4°C the eggs took 8 days to hatch. At room temperatures the mean durations of the larval instars were 7.5 days in instar I, 10.0 days in II, 14.0 days in III and 45.0 days in instar IV. The reported period between generations was 5 to 6 months. Larvae survived temperatures below freezing for at least 48 h and also survived desiccation. The species probably normally overwinters in the larval stage. Goldie-Smith (1989a) reported clutch sizes of 15 to 55 eggs, but typically 20–40, in November. She confirmed that at room temperatures the incubation of the egg took only 5 days. She recorded an aerial swarm of males on a mild and sunny January 24th.

Distribution. Throughout mainland Britain and Ireland, but commonest to the south of a line joining the Humber to the Mersey. *D. autumnalis* is the most frequently encountered species of *Dixella* in lowland England (Peach 1984).

Dixella filicornis

Larvae recorded January–November; adults January–March, June–November. Larval habitats include rushes bordering a eutrophic lake fringed with trees, the edge of a small loch (Morgan & Waddell 1961), a hydrosere woodland swamp, trailing *Deschampsia caespitosa* in waters shaded by alders (Fowler 1984a), marginal slack water in an unshaded, shallow, stony roadside ditch (Fowler 1987b), a marshy pool with watercress, a small woodland stream, a flooded limestone quarry and a drainage culvert (Goldie-Smith 1990b).

In Perthshire, adults entered an emergence trap in the early morning (Morgan & Waddell 1961).

Distribution. An uncommon species found at scattered localities throughout the lowlands of mainland Britain and Ireland, but mainly south of a line joining the Mersey and the Humber.

Dixella graeca

Larvae recorded May, August and October; adults August–September. Larval habitats include reed swamps and marginal vegetation in a small pond.

Observations on the immature stages and courtship behaviour of the adults are presented by Goldie-Smith (1993). Typically there are 30–40 eggs per clutch, but up to 83 have been recorded.

Distribution. So far, only known from three sites, in Cambridgeshire, Suffolk and East Sussex (Disney 1992; Goldie-Smith 1993).

Dixella martinii

Larvae recorded February–December; adults March–December. Larvae are found in beds of rushes, reeds, sedges and grasses in waters ranging from acid to alkaline, from oligotrophic to eutrophic, and in habitats that range from wet seepages, dune slacks, water-filled tractor ruts, rich fens, ponds and watercress beds, to the margins of ditches, lakes, streams and rivers. In the absence of other species (e.g. on Shetland) it will extend its habitat range into hill-streams (Fowler 1984b). Associated plants include *Phragmites, Carex, Juncus, Schoenus, Typha, Glyceria, Molinia, Potentilla, Oenanthe, Myrica* and *Menyanthes*. Adults have been recorded in a cave and a mine (Goldie-Smith 1990b).

Goldie-Smith & Thorpe (1991) reported that the incubation of the egg lasts about five days at room temperature. There are up to 100 or more eggs per clutch, which is laid in the meniscus or just above.

Distribution. Throughout the British Isles. Recorded from several offshore islands, such as Shetland, Foula, Fair Isle and Bardsey (Fowler 1984b, 1987a, 1989).

Dixella obscura

Larvae recorded April–September; adults March–October. The larval habitat is typically sedge swamps, especially *Carex rostrata*, in calcareous waters to more acid waters but typically with a trace of calcium, as indicated by the present of at least one species of snail. *D. obscura* possibly requires a calcium concentration greater than 1 mg per litre. The habitats include seepages, ponds and the edges of slow-flowing streams.

Distribution. Scattered localities in mainland Britain north of a line from the Mersey to the Humber.

Dixella serotina

Larvae recorded March–May, August–October and December; adults January–December. The larvae, which are particularly partial to lowland fens, typically occur in reed (*Phragmites australis*) swamps, but are also found in beds of sedges (especially *Carex riparia*) and hydroseres with reeds, rushes, grasses and iris. Occurs in a calcareous valley fen with *Carex elata*. Fowler (1984a) reports on a typical habitat for this species. The mortality of larvae due to surfactant pollutants is reported by Fowler, Withers & Dewhurst (1997).

Distribution. Lowlands of the Midlands, Hampshire and East Anglia, the western fringes and southern half of Wales, and Ireland.

THAUMALEIDAE (Trickle Midges)

Recent introductions to this family are provided by Wagner (1997a,c). The adults of the three recorded British species of trickle midges were covered by Edwards (1929). By contrast, the European fauna includes nearly 100 species, the majority of which are in the Alps. Vaillant provides keys to males (1977) and to females (1981). However, new species are regularly being added. Those previously unrecorded species most likely to turn up in Britain are covered by Wagner (1997a). The larva and pupa of at least one of the British species has been described (Saunders 1923; Bertrand 1948; Nielsen, Ringdahl & Tuxen 1954; Sinclair 1996), but with some confusions (see below). The egg and earliest instars of a second species were described by Mandaron (1963).

CHECKLIST OF BRITISH THAUMALEIDAE

Genus *Thaumalea* Ruthé, 1831
Orphnephila Haliday, 1832
1. *testacea* Ruthé, 1831
devia (Haliday, 1832)
2. *truncata* Edwards, 1929
3. *verralli* Edwards, 1929

A reference collection of slide-mounted British Thaumaleidae has been deposited in the Zoology Museum of Cambridge University.

MORPHOLOGY OF THE EGGS OF *THAUMALEA*

Edwards (1929) reported that the eggs have 5 to 8 deep parallel grooves running the entire length of the dorsal surface and such eggs were first illustrated by Mandaron (1963). Eggs of *T. verralli* are shown in Fig. 34.

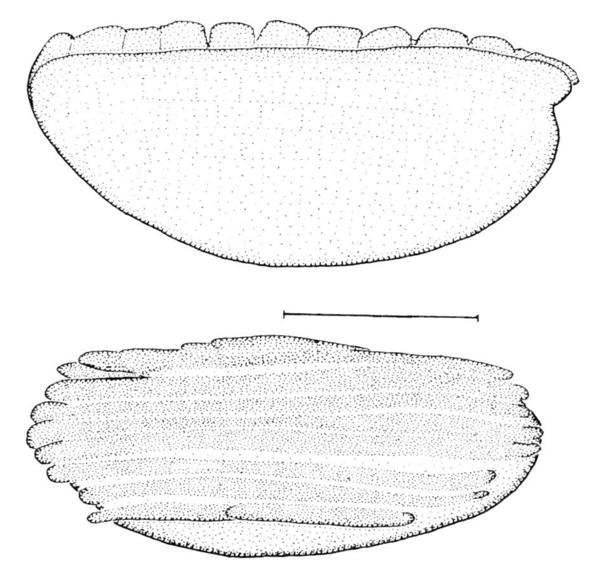

Fig. 34. Eggs from a gravid female of *Thaumalea verralli*, in side view (above) and dorsal view (below). (Scale bar = 0.1 mm).

MORPHOLOGY OF THE LARVAE OF *THAUMALEA*

The larvae (Fig. 35) are remarkable creatures. Saunders (1923) noted that each body segment bears a peculiar dorsal "saddle" of reticulated chitin and suggested that histological preparations of fresh specimens would be necessary to reveal their true nature. Indeed, in slide-mounts under the light microscope these saddles are colourless and difficult to observe, except in a moulted pelt (exuviae). However, their structure is admirably revealed with the scanning electron microscope (Plate 8). Whether these elegant structures function as a plastron is not known. They certainly enable the larva to suspend itself horizontally from the surface film in a manner reminiscent of a *Dixa* larva.

The amplified description of the putative larva of *Thaumalea testacea* by Saunders (1923) was based on a sample of the specimens collected by Thienemann and described and attributed to this species by him (Thienemann 1910). However, Edwards (1929) stated "but its specific identity is questionable". I endorse this opinion and elaborate below. The larva of *T. verralli* was described by Bertrand (1948) and also by Nielsen, Ringdahl & Tuxen (1954). However, the supposed difference between this species and the larva described by Thienemann and Saunders, as highlighted by Nielsen *et al.*, was based on their misinterpretation of a text-figure by Saunders (1923). Furthermore the

Plate 8. Detail of the reticulated "saddle" on abdominal segment 8 of the larva of *Thaumalea verralli*, viewed under a scanning electron microscope. (Photograph by R. H. L. Disney and W. M. Lee).

excellent text-figures of the head capsule given by Saunders (his Figure 1 on page 634) is evidently that of *T. verralli*, whose larvae I have identified by my own successful rearings of adults of this species from the final larval instars (see below). Thienemann's figure of the head capsule is of the same species. The larva of the Holarctic *T. verralli* has been more recently described by Sinclair (1996) and his Figs 28 and 29 are similar to my Fig. 42A, which is drawn from the empty head capsule of the larval pelt from which an adult male subsequently emerged.

Either Thienemann's misidentification was based on an erroneous association of adults and larvae through having collected larvae and adults of two species at the same place on the same day (see the "caution" below), or he had misidentified the associated adults. In view of the fact that *T. verralli* was not described and distinguished from *T. testacea* until 1929, the latter must be a distinct possibility. It is also worthy of note that Thienemann attributed the authorship of the specific epithet "*testacea*" to Macquart, not Ruthé. Lindner

(1930) cites Macquart's name only as a possible synonym of Ruthé's species.

Mandaron (1963) described the first-instar larva and the youngest instars of *T. testacea*. From my own rearing of larvae, and saving the moulted pelts, data have been obtained on differences between later instars. For example, the lateral processes of the median protuberance (m in Fig. 36) of the head of *T. verralli* clearly vary with the instar. Thus in one sequence they are relatively short, in comparison with the median process, in the last instar (L); they are relatively longer in the penultimate instar (L–1), and are subequal in length to the median process in the previous instar (L–2) (Fig. 42B). These differences, along with a remarkable degree of intraspecific variation, have greatly complicated recognition of the species in the larval stages.

An added complication is that the number of larval instars is variable. While there are four in Dixidae, in the supposedly closely related Simuliidae (see page 8) there are 6 to 9 instars (Crosskey 1990). Mandaron (1963) reported that the number of instars in *T. testacea* is not fixed, but in his laboratory rearings ranged from 15 to 20, with the number varying according to the availability of food. I present the following results based on a study of larvae and the pelts of known instars obtained in my rearing experiments. These are either last instars (L) – usually recognisable in slide mounts by the presence of the developing pupal respiratory trumpets within, the penultimate instars (L–1), or the instar before the penultimate (L–2), or larvae smaller than the smallest recorded L–2 larva. As the results suggest that there are normally fewer instars than the numbers reported by Mandaron, I summarise my observations below.

A pilot study was carried out to select a suitable indicator of age. The anterior spiracle (Fig. 35) proved to be the most useful feature, with at least one being clearly visible in slide-mounted larval pelts. The number of rays (radial bands) at the rim was scored and the maximum width of the spiracle was measured (to the nearest 5 µm) for each larva or pelt whose instar was known. While there is a linear relationship between the number of rays and the width of the spiracle, the ranges for the number of rays in each instar overlap with those in adjacent instars. It was therefore decided to base further analysis on the maximum widths alone. In *T. verralli*, the widths of the spiracles of larvae collected in the field ranged from 10 to 85 µm. Mandaron (1963) reported that the first instar lacks the anterior spiracles and the posterior pair are not united by a common spiracular plate. My samples, therefore, did not include first-instar larvae. The results for larvae of known instars, and for those of unknown instars that moulted in the laboratory, are given in Table 1.

For *T. verralli* larvae, the median maximum width of the spiracle in the last instar (L) was 70 µm (range 60–85, n = 38). For L–1 instars the median was

Table 1. Frequencies of the maximum width (μm) of the anterior spiracle in larvae of *Thaumalea verralli* collected in the field and in the laboratory. L = last instar; L–1 = penultimate instar; L–2 = instar before penultimate instar; L–? = unknown instar, measured at time of the moult (see the text).

	Larval instar			
Max. width	L	L–1	L–2	L–?
10	–	–	–	–
15	–	–	–	1
20	–	–	–	9
25	–	–	–	12
30	–	–	1	3
35	–	–	–	4
40	–	–	–	3
45	–	–	2	–
50	–	5	–	5
55	–	5	2	1
60	3	7	1	2
65	9	6	2	1
70	12	4	–	–
75	3	3	–	–
80	9	–	–	–
85	2	–	–	–

60 μm (range 50–75, n = 30) and for L–2 instars it was 55 μm (range 30–65, n = 8). For larvae of unknown instars (L–?) that moulted in the laboratory, the widths at the time of moult (i.e. width of the spiracles on the pelt) were also measured (Table 1). While I have more limited measurements for *T. testacea* and *T. truncata*, these are entirely consistent with the results shown for *T. verralli*.

Of the L–? records, most of those larvae whose spiracle widths exceeded 25 μm are probably L–2 larvae along with at least some of those measuring 25 μm. If one hypothesises that in a larger sample of established L–2 larvae they would have spiracle widths ranging from 25 to 65 μm, then L–3 larvae possibly range from *ca.* 20 to *ca.* 30 μm or from 5 to *ca.* 30 μm. The former seems less likely given the greater range in L–2 instars compared with L and L–1 instars. If this trend applies to the L–3 instar, then a range of 5 to *ca.* 30 μm would be consistent with it. I conclude that in my samples there is a minimum of four larval instars, but there is a possibility of five. Further rearings need to concentrate on larvae whose anterior spiracle is less than 30 μm in width. As first-instar larvae would score 0 (and are just under 1 mm in length: Popham 1952; Mandaron 1963), then my data based on these field samples suggest that 5 to 8 instars might be nearer the norm under field conditions in the months of June to August.

The parts of the larva are illustrated in Figs 35 and 36. The protuberances of the head capsule (Fig. 36A) would seem to be the most taxonomically useful features despite being somewhat variable. Indeed they are more distinctive in the different species in younger larvae than in the last instars. The protuberances on the sides of the head are absent from the first instar but the median protuberance is present and functions as an egg tooth at the time of eclosion from the egg (Mandaron 1963).

CAUTION. It is a common occurrence to find the larvae of two species coexisting in the same place at the same time, even when the adults of only one species are caught at the time of sampling. I have also recorded larvae of all three species coexisting. Indeed, the separation of the larvae of *T. testacea* and *T. truncata* remains unsatisfactory. While I have sampled populations of *T. testacea* coexisting with *T. verralli* only, I have not yet encountered larval populations of *T. truncata* that were not coexisting with both of the other species. Consequently I remain uncertain as to how to distinguish young larvae of *T. testacea* and *T. truncata*.

A key to the larvae of *Thaumalea* begins on p. 98.

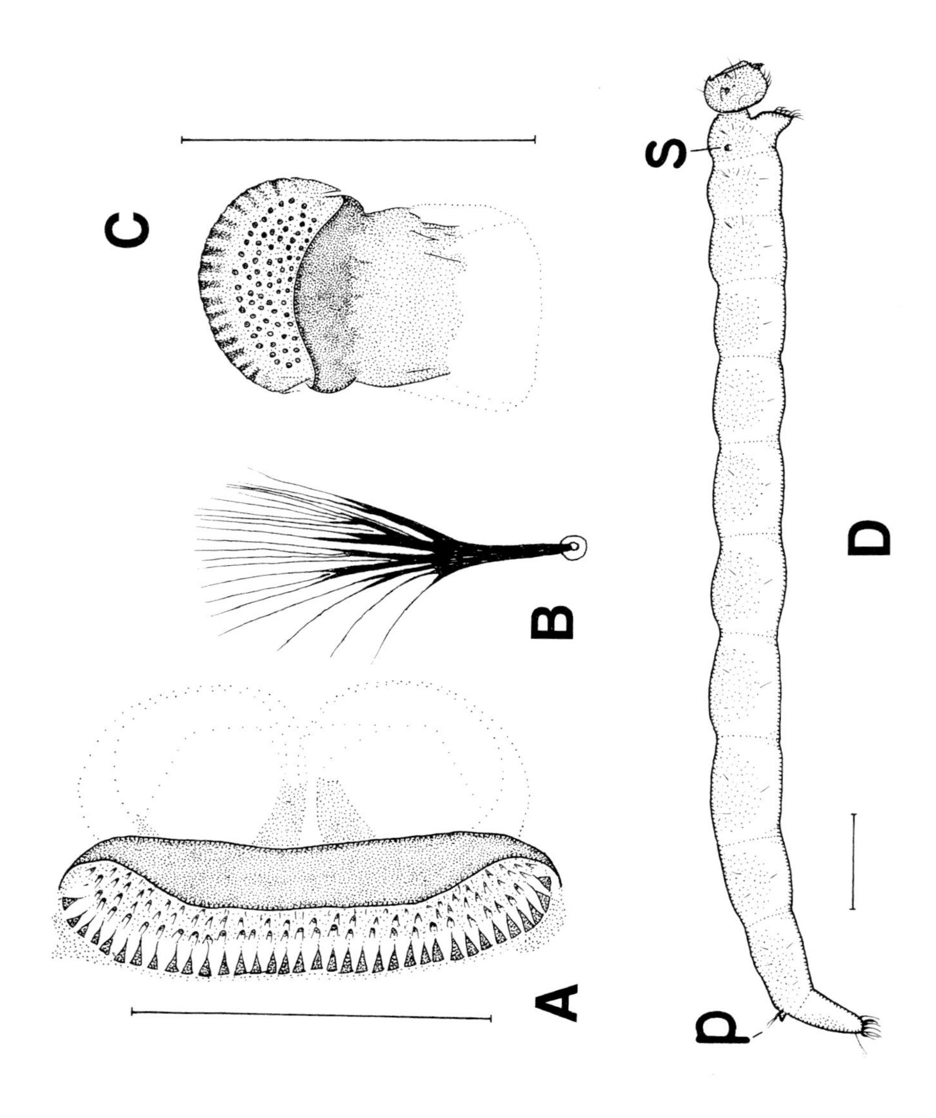
A
B
C
D
S
P

Fig. 35. *(On facing page)*. Larva of *Thaumalea testacea* (**D**) with enlarged details above: **A**, posterior spiracular plate; **B**, an abdominal branched bristle; **C**, an anterior spiracle (right). **D**, larva viewed from the right-hand side: p = bristles on the tip of the procercus; s = anterior spiracle. (Scale bars for **A–C** = 0.1 mm; scale bar for **D** = 1.0 mm).

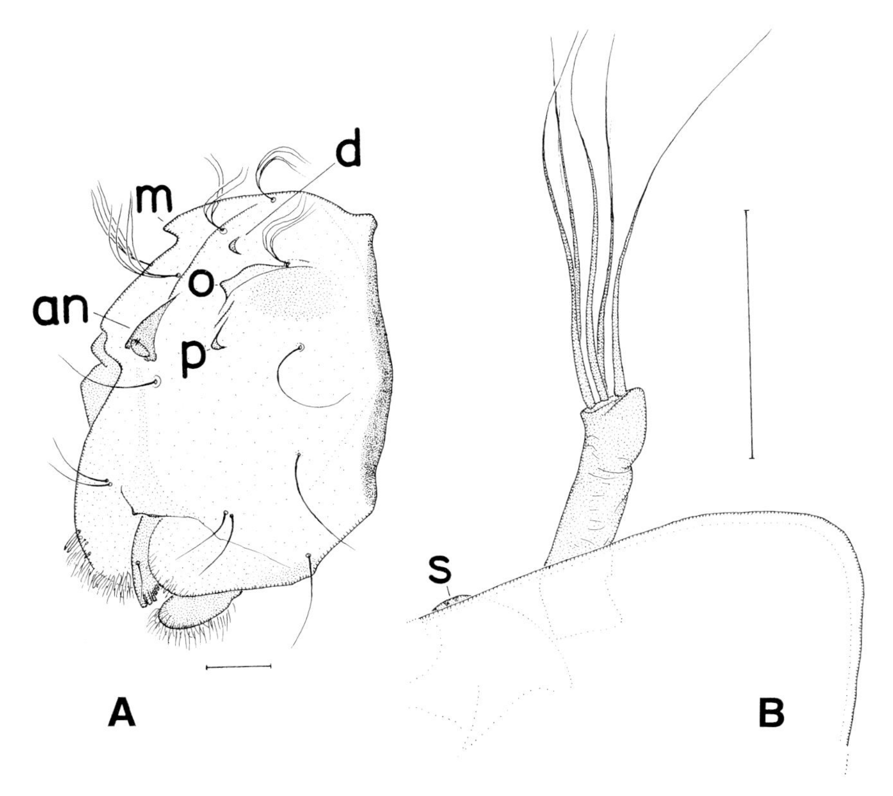

Fig. 36. Details of the head capsule (**A**) and left procercus (**B**) of the last-instar larva of *Thaumalea testacea*, viewed from the left-hand side. On the head capsule: an = antenna; d, m, o and p = dorsolateral, median, ocular and postantennal protuberances, respectively. Near the procercus: s = left margin of posterior spiracular plate. (Scale bars = 0.1 mm).

KEY TO LARVAE OF *THAUMALEA*

4 Lateral protuberances of head capsule and anterior spiracles absent. The two posterior spiracles are completely separated from each other—
1st-instar larvae of **Thaumalea** species

[NOTE: 1st-instar larvae of the three British species cannot be distinguished in our current state of knowledge.]

— Lateral protuberances of head capsule and anterior spiracles present (Figs 35, 36). Posterior spiracles united (Fig. 35A)— Later instars, **2**

2 Median protuberance of head capsule typically comprising three tapered processes (Figs 37m, 38m), but there may be supplementary shorter processes and/or one lateral process may be abbreviated (e.g. Figs 37A, 38B)— **Thaumalea verralli** Edwards (part)

— Median protuberance less tapered and with a more rounded tip and either without lateral processes or if these are present they are relatively short and with more rounded tips (e.g. Figs 41, 42A)— **3**

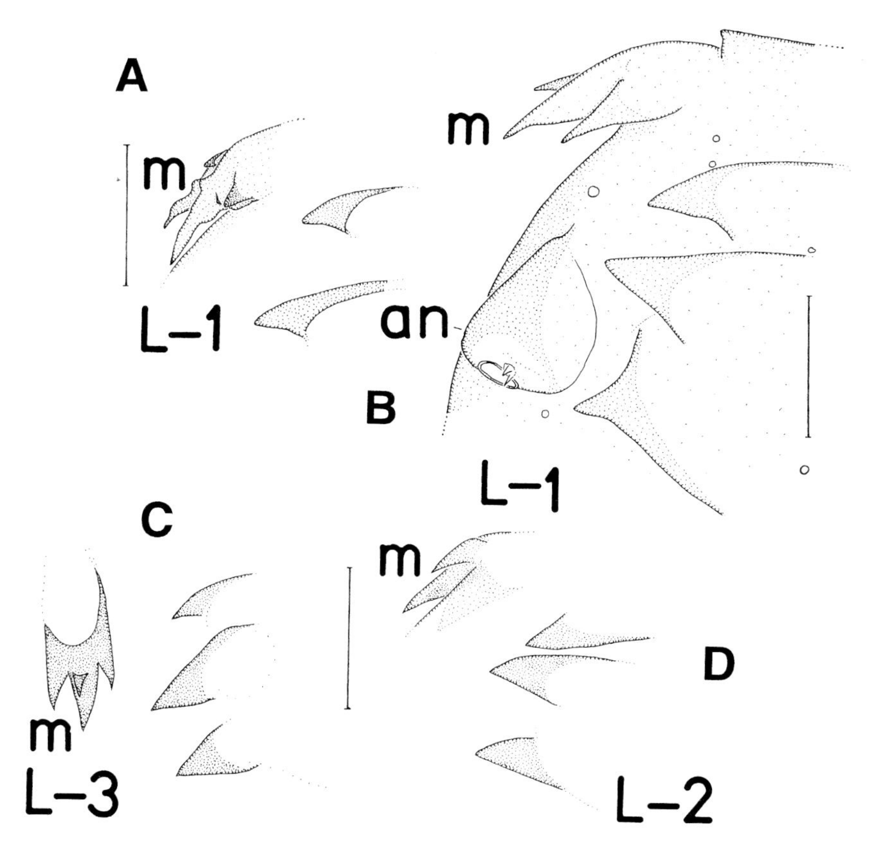

Fig. 37. Drawings of the median process (m) and left lateral protuberances (d, o, p; see Fig. 36) on the head of larvae of *Thaumalea verralli*. **A**, penultimate instar (L–1). **B**, penultimate instar (L–1), left face of head, including the antenna (an). **C, D**, L–3 and L–2 instars of the same larva, with median process (m) viewed from above in **C**. (Scale bars = 0.1 mm).

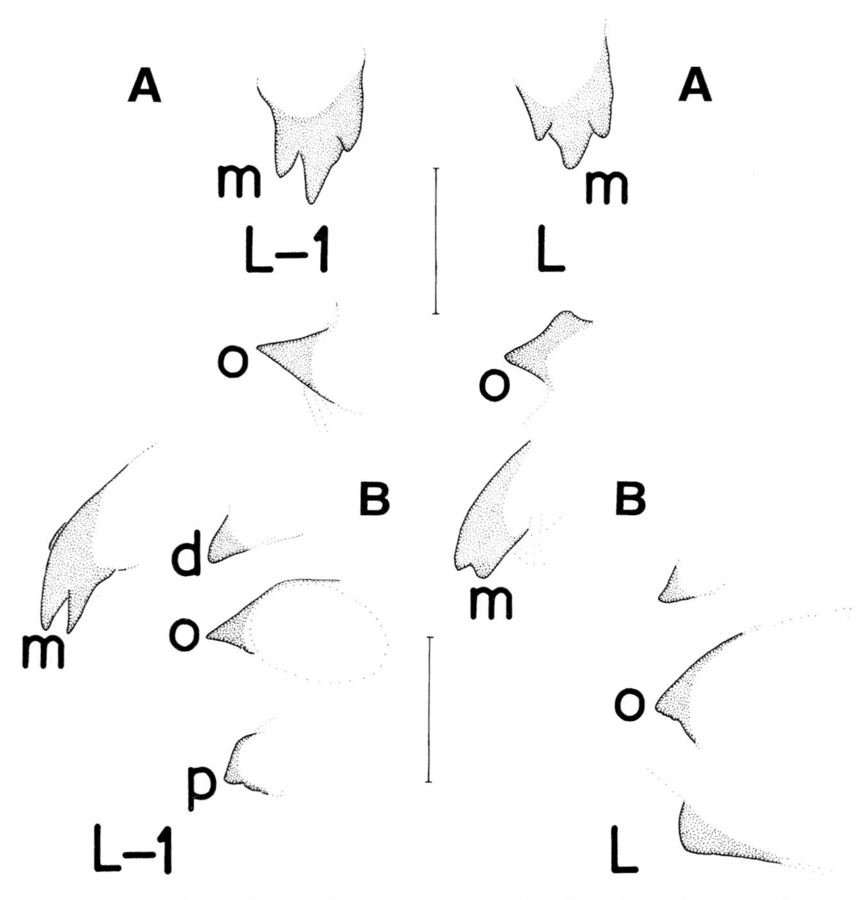

Fig. 38. Drawings of the median process (m) and left lateral protuberances (d, o, p) on the head of larvae of *Thaumalea verralli*. **A**, median process viewed from above and left faces of the ocular protuberance of the same larva in the penultimate (L–1) and last (L) instars. **B**, left faces of the median process and protuberances in another larva in the penultimate (L–1) and last (L) instars. (Scale bars = 0.1 mm).

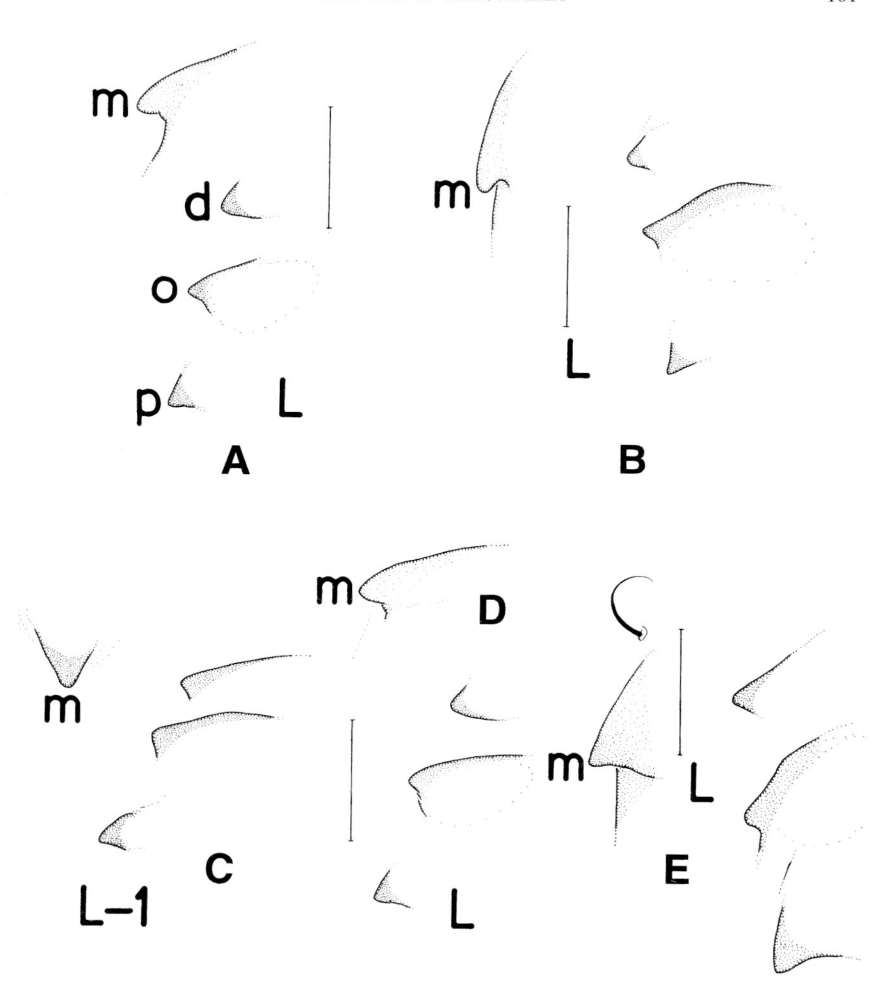

Fig. 39. Drawings of the median process (m) and left lateral protuberances (d, o, p) on the head of larvae of *Thaumalea testacea*. **A, B,** last-instar larvae (L). **C, D,** penultimate instar (L–1) and last instar (L) of the same larva, with median process (m) viewed from above in L–1 and from the left in L. **E**, median process and lateral protuberances in a last-instar larva (L) from which an adult female developed. (Scale bars = 0.1 mm).

3 Median protuberance of head capsule entirely lacking lateral processes (e.g. Figs 39 and 40)— **4**

— Median protuberance with at least short lateral processes (e.g. Figs 41 and 42A) or bifid at tip— **5**

4 Lateral protuberances of head capsule relatively large in last instars (Figs 36, 39)— **Thaumalea testacea** Ruthé

— Lateral protuberances of head relatively small in last instars (Fig. 40)— **Thaumalea truncata** Edwards

[NOTE: In our current state of knowledge, it is not always possible to separate some larvae of these two species, especially some of the earlier instars].

5 In final instars the lateral processes of median protuberance of head originate nearer base and lateral protuberances are relatively large (Fig. 42A)— **Thaumalea verralli** Edwards (part)

— In final instars the lateral processes of median protuberance originate further from base and lateral protuberances are relatively small (Fig. 41)— **RETURN TO COUPLET 4**

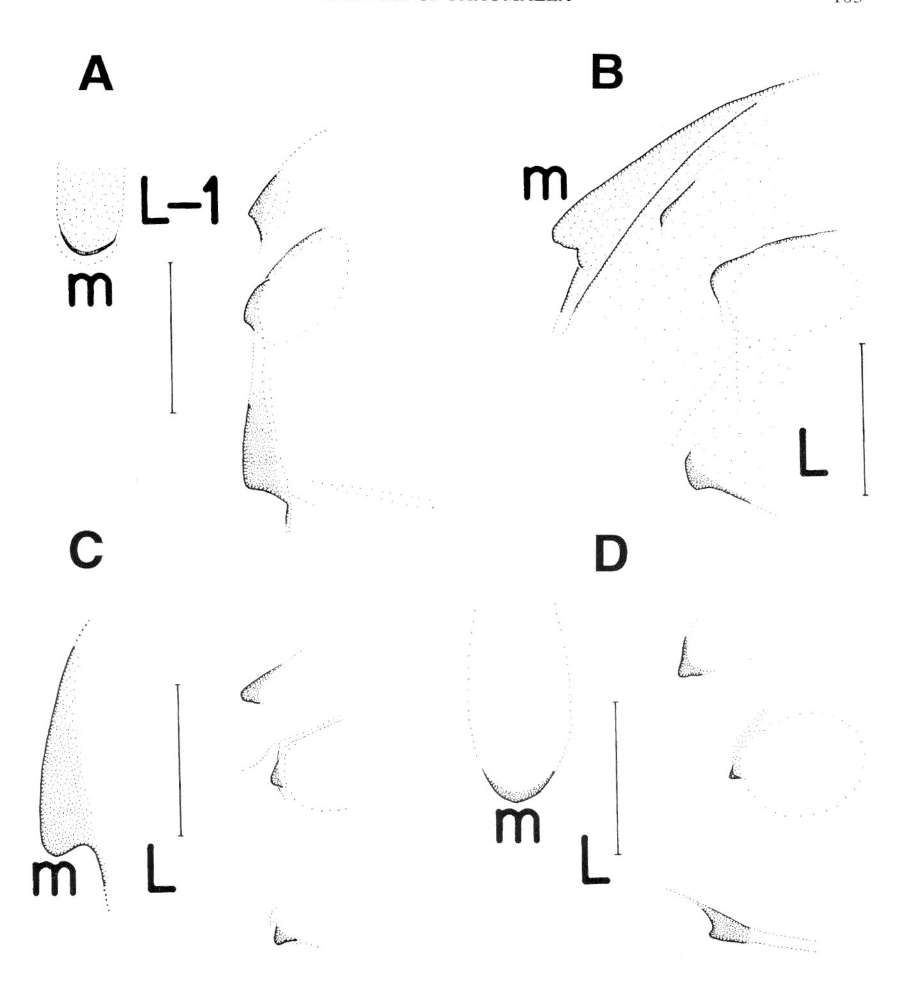

Fig. 40. Drawings of the median processes (m) and left lateral protuberances on the head of larvae of *Thaumalea truncata*. **A, B**, penultimate (L–1) and last (L) instar of same larva. **C**, last instar (L) from which an adult male developed. **D**, last instar (L). In **A** and **D** the median process is viewed from above. (Scale bars = 0.1 mm).

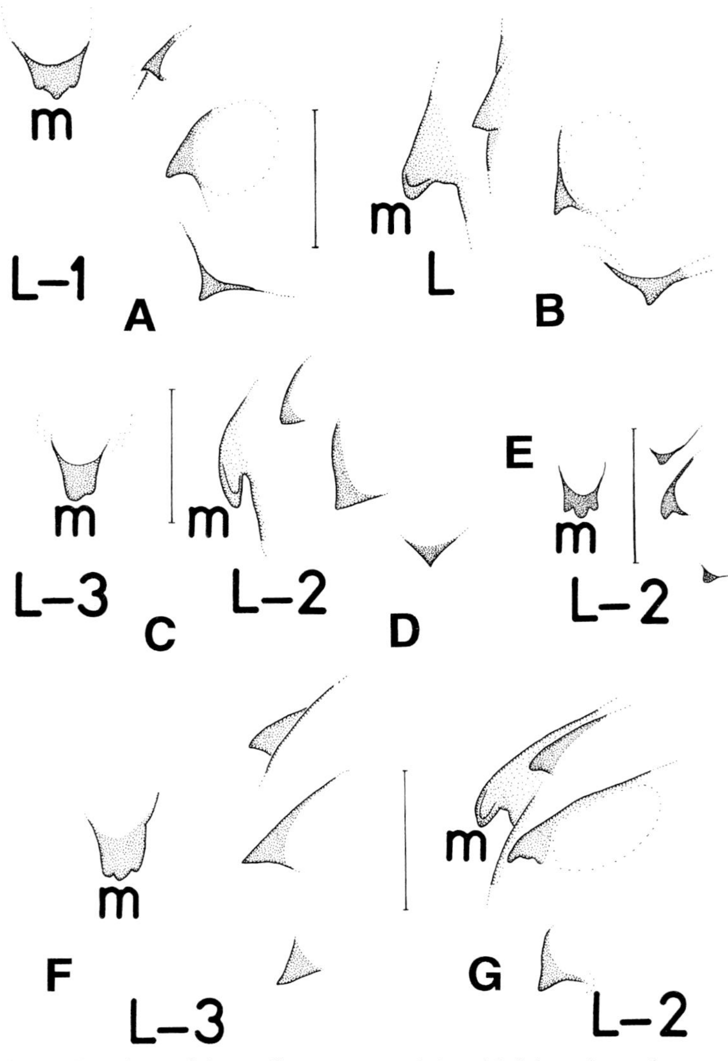

Fig. 41. Drawings of the median processes (m) and left lateral protuberances (d, o and p in Fig. 36) on the head of putative larvae of *Thaumalea testacea*. **A** and **B**, **C** and **D**, **F** and **G** are drawings from the same larvae, respectively, in last (L), penultimate (L–1), L–2 and L–3 instars as indicated. In **A, C, E, F** the median process is viewed from above. (Scale bars = 0.1 mm).

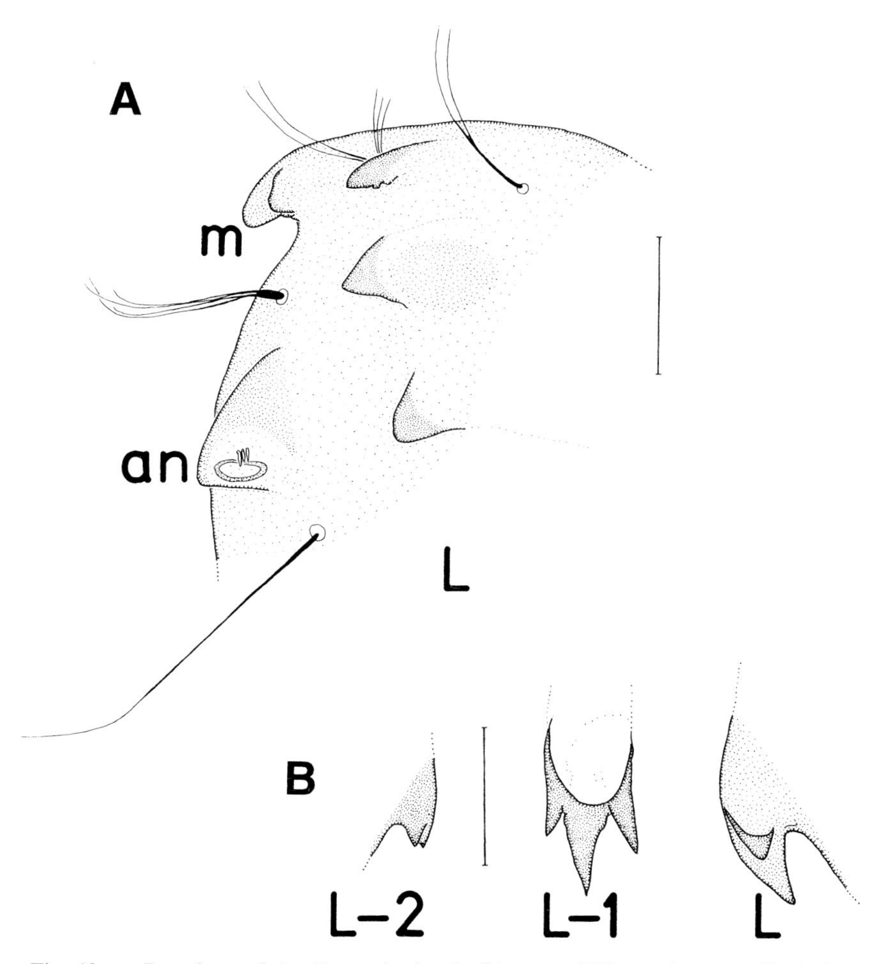

Fig. 42. Drawings of details on the head of larvae of *Thaumalea verralli*. **A**, last instar, from which an adult male developed, showing left face of the anterior portion of the top half of the head capsule (an = antenna; m = median process). **B**, median protuberance of head in last instar (L) viewed from the left side, in penultimate instar (L–1) from above, and in the previous instar (L–2) from the right side, all in the same larva. (Scale bars = 0.1 mm).

PUPAE OF *THAUMALEA*

The pupa of *Thaumalea testacea* is shown below, in Plate 9. Live pupae are best identified by rearing out the adult. Mature preserved pupae can be identified by dissecting out the abdominal terminalia. With the limited material available and the evident intraspecific variation, I have not attempted to give a key to the species.

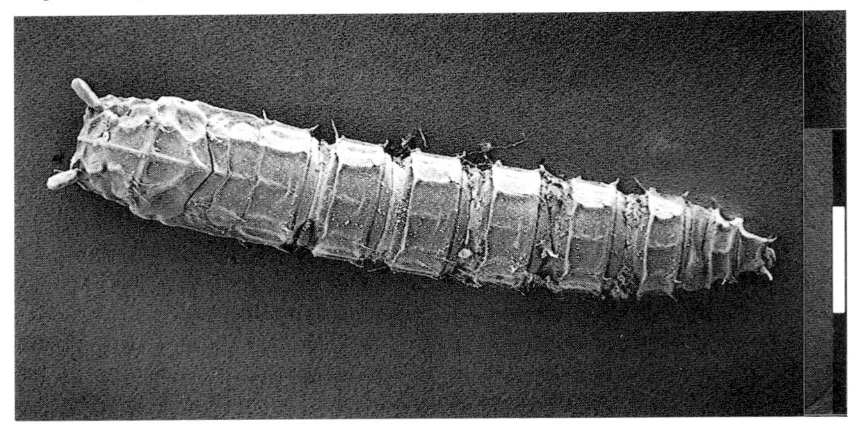

Plate 9. Pupa of *Thaumalea testacea*: SEM photograph (by R. H. L. Disney and W. M. Lee). (The white scale bar is 1 mm).

ADULTS OF *THAUMALEA*

Trickle midges (Fig. 43) are immediately distinguished from meniscus midges (Fig. 20) by their shorter legs and antennae. They most resemble blackflies (Simuliidae), but in the outer half of the wing the latter have 3–4 anterior longitudinal veins reaching the wing margin well in front of the wing tip. Thaumaleidae have only 1–2 anterior longitudinal veins (Plate 10) and furthermore they lack the large anal lobe at the base of the wing, which is characteristic of blackflies.

Both sexes of the three British species were covered by Edwards (1929), who worked with pinned material. The following key is based on slide-mounted specimens. The adults vary in size and hence the wings of the three species vary in size; consequently they exhibit some variations due to the effects of allometric growth. The abdominal terminalia, therefore, have been used to distinguish the species in the key given on page 110.

Each male coxite bears two processes on its inner face. The larger may be long and complex while the smaller one is relatively inconspicuous. Edwards referred to the larger process as a paramere, but it is now thought that the smaller process is homologous with the paramere in related families. The larger, taxonomically useful process is now referred to as the coxal (or gonocoxal) blade (e.g. Sinclair 1996). That of *T. testacea* is not only complex (Fig. 45D) but its two most apical points are variable in length (Schmid 1951). The posterolateral processes of the female abdominal tergite 9 are likewise somewhat variable in *T. verralli* (Bertrand 1948, and Fig. 46A).

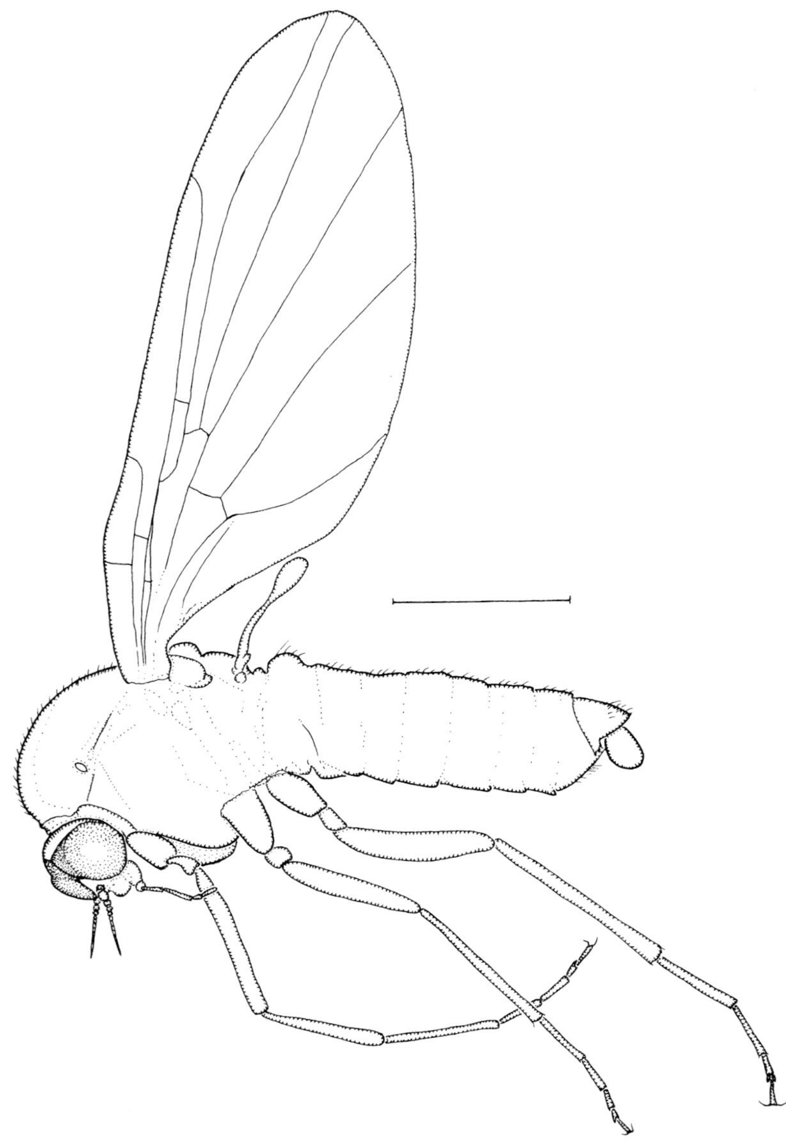

Fig. 43. An adult female trickle midge, *Thaumalea verralli*, viewed from the left-hand side. (Scale bar = 1 mm).

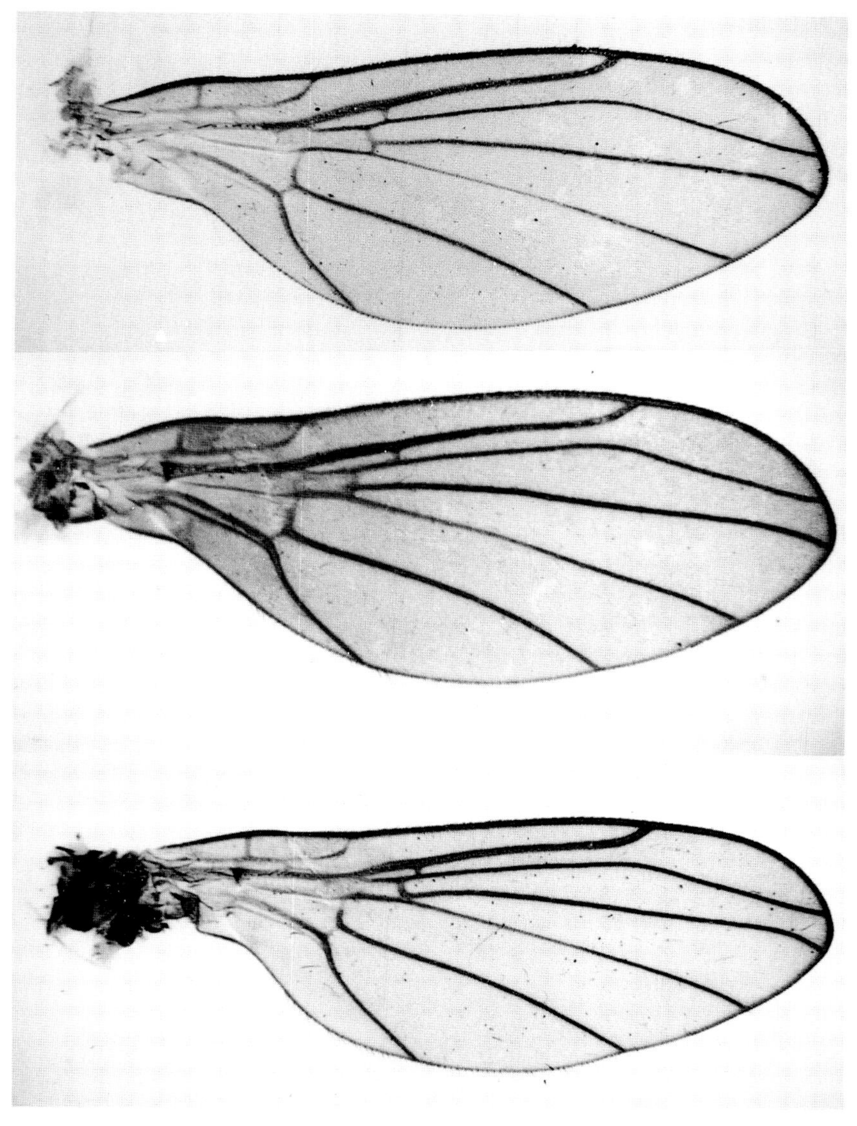

Plate 10. Wings of British Thaumaleidae. In each case the right wing of a male is illustrated. Top: *Thaumalea testacea*. Middle: *T. truncata*. Bottom: *T. verralli*. (Photographs by the author).

KEY TO ADULTS OF *THAUMALEA*

1 Abdomen ending in a pair of claspers each consisting of a coxite and style (Figs 44A, C, 45C and 46C)— MALES, **2**

— Abdomen ending in a pair of rounded cerci (Figs 46A, 47)— FEMALES, **4**

2 Styles twisted in middle and ending in two dark terminal teeth plus a subterminal tooth (Fig. 44A, B). Coxal blades (see Fig. 44A, C) as in Fig. 44D— **Thaumalea truncata** Edwards MALE

— Styles and coxal blades otherwise— **3**

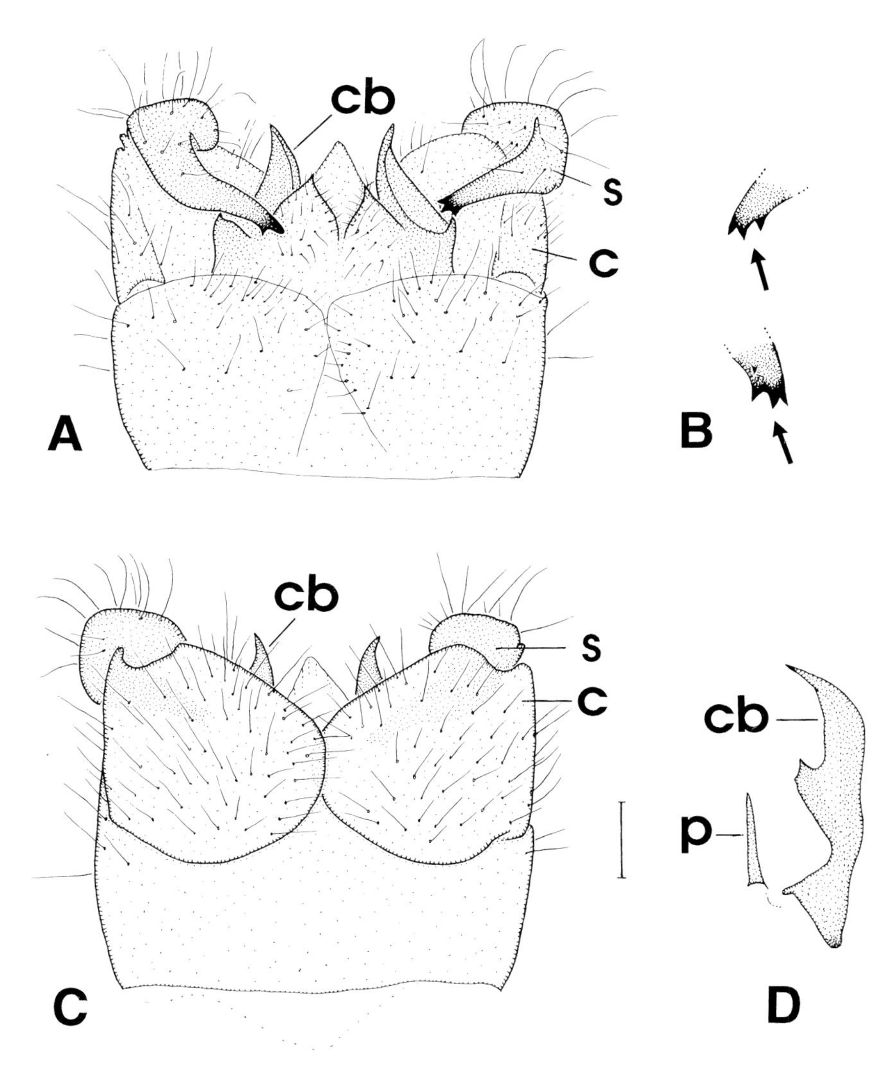

Fig. 44. Abdominal terminalia of an adult male *Thaumalea truncata*, viewed from above (**A**) and below (**C**); ↗s styles, ↗c coxa. **B**, ↗ tips of styles drawn from two different angles. **D**, Views of the right paramere (↗p) and coxal blade (↗cb). (Scale bar = 0.1 mm).

3 Coxal blades relatively short but complex (Fig. 45D). Styles relatively short (Fig. 45C). The two small projections at rear of tergite 9 close together in middle (Fig. 45B)— **Thaumalea testacea** Ruthé
MALE

— Coxal blades long but relatively simple and styles relatively long (Fig. 46C). The two small projections at rear of abdominal tergite 9 are widely separated (Fig. 46B)— **Thaumalea verralli** Edwards
MALE

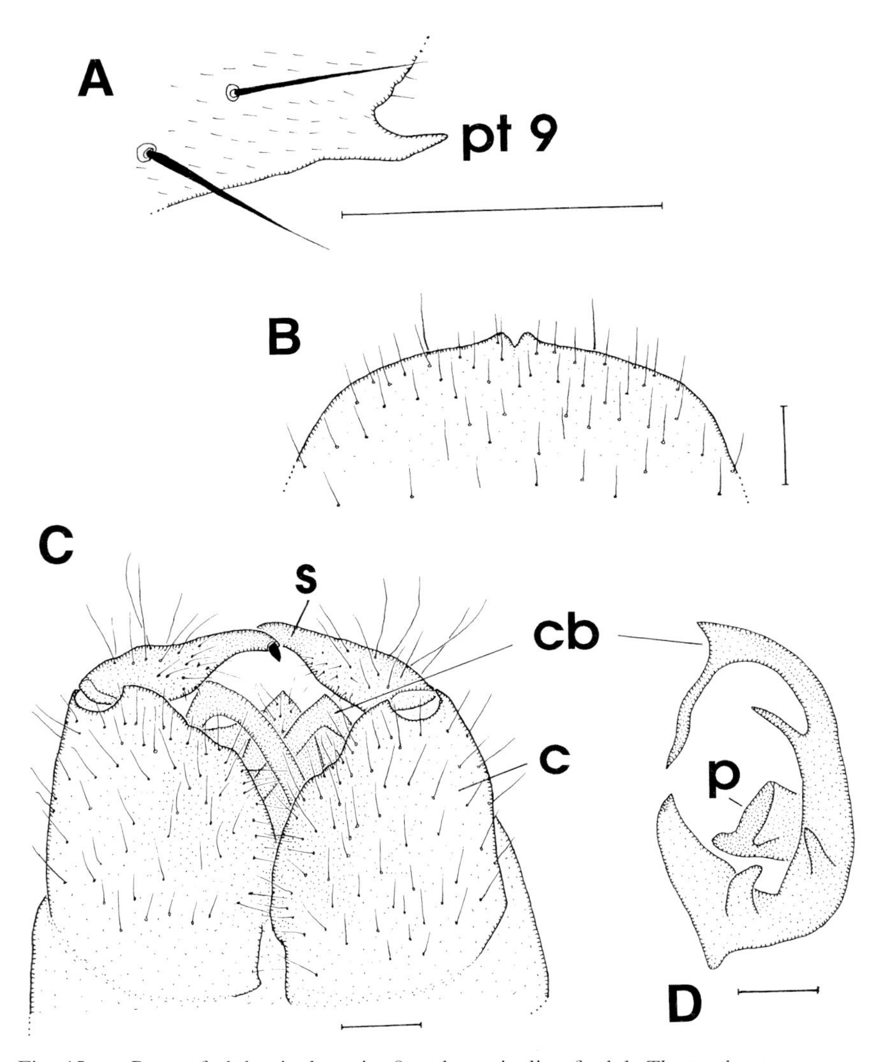

Fig. 45. Parts of abdominal tergite 9 and terminalia of adult *Thaumalea testacea*. **A**, posterolateral process (↗pt9) on left side of female abdominal tergite 9. **B**, posterior border of male abdominal tergite 9. **C**, male terminalia viewed from below (↗cb coxal blade; ↗c coxa; ↗s style). **D**, male right-hand coxal blade (↗cb) and paramere (↗p). (Scale bars = 0.1 mm)

4 (1) Each posterolateral region of abdominal tergite 9, below the bases of the cerci, with a large untapered process whose hind margin has a variable number of teeth (Fig. 46a, ↗pt9)— **Thaumalea verralli** Edwards FEMALE

— No such processes, any that are present being smaller and tapered to a single point (e.g. Fig. 45A, ↗pt9)— **5**

Fig. 46. (*On facing page*). Parts of abdominal terminalia (segment 9 and beyond) in adult *Thaumalea verralli*. **A**, female abdominal tergite 9 with the left cercus (↗cer) and the untapered posterolateral process (↗pt9) with a variable number of teeth (viewed from left side); six variants of the tip of the posterolateral process are shown to the right and below the drawing. **B**, male abdominal tergite 9, drawn with the rear margin to the left. **C**, male terminalia viewed from below (↗c coxa; ↗cb coxal blade; ↗s style). (Scale bars = 0.1 mm).

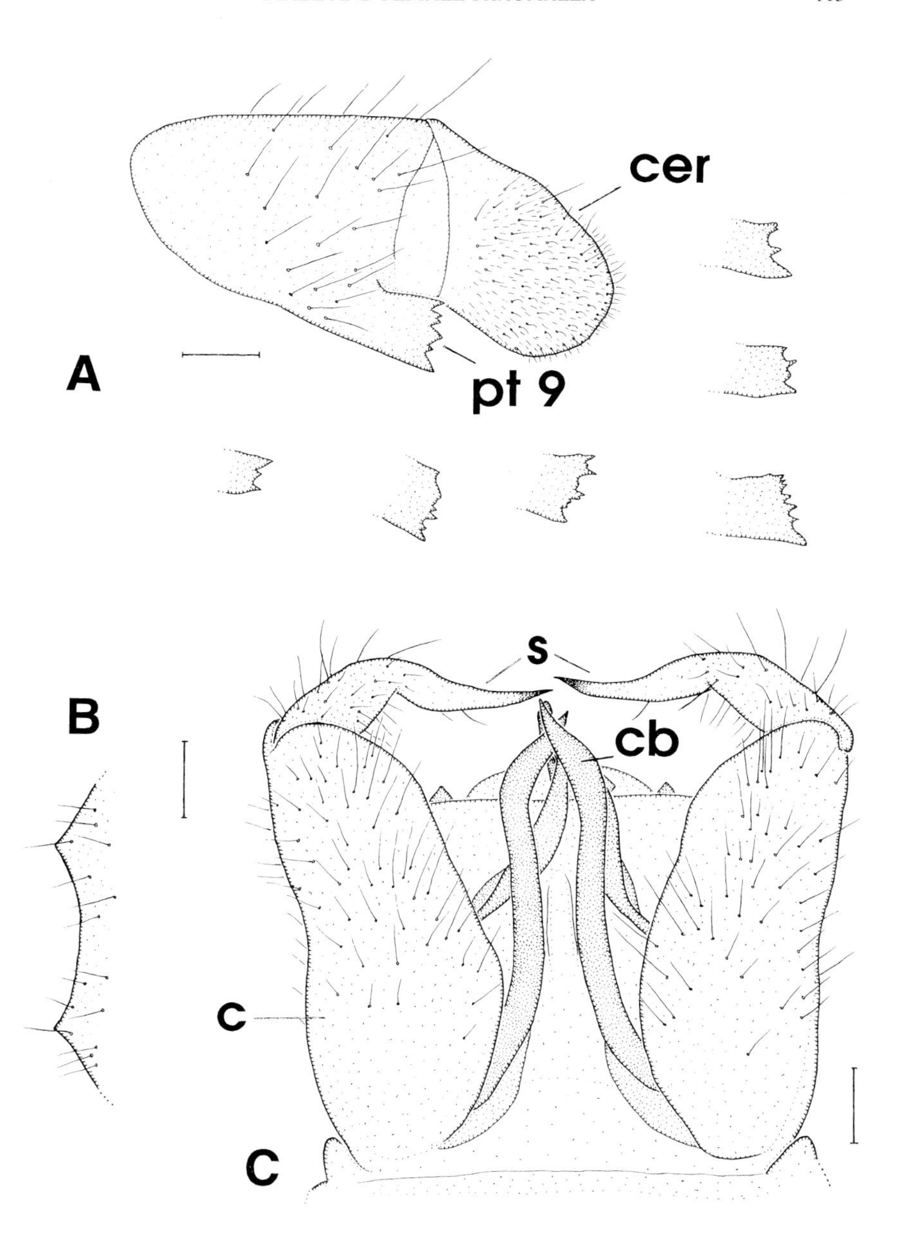
cer
A
pt 9
B
s
cb
c
C

5 The tips of the pair of processes at rear margin of abdominal sternum 8 (= "hypogynial valves") are pointed (Fig. 47A, ➚pas). A small tapered posterolateral process on each side of abdominal tergite 9 (Figs 45A and 47A, ➚pt9)— **Thaumalea testacea** Ruthé
FEMALE

— The tips of the processes at rear margin of sternum 8 are rounded (Fig. 47B, ➚pas). No processes on posterolateral regions of abdominal tergite 9 (Fig. 47B, ➚pt9)— **Thaumalea truncata** Edwards
FEMALE

Fig. 47. (*On facing page*). Parts of abdominal terminalia (segments 8 and beyond), with the cercus (➚cer) in adult females of *Thaumalea*, viewed from the left-hand side. **A**, *T. testacea* abdominal segment 8 with pointed posterior abdominal process (pas) on sternum 8 (➚s8), and tapered posterolateral process on tergite 9 (➚pt9). **B**, *T. truncata* abdominal segment 8 with a rounded posterior abdominal process (➚pas) on sternum 8 (➚s8), and without a posterolateral process on tergite 9 (➚pt9). (Scale bars = 0.1 mm).

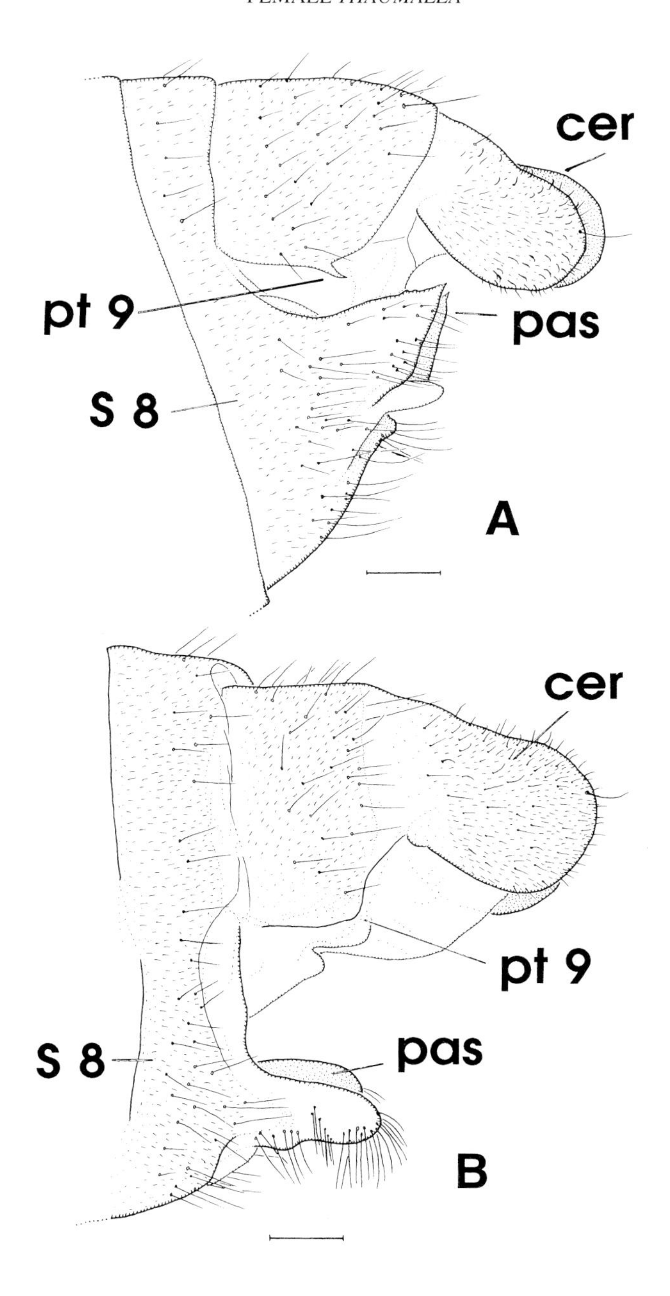
cer
pt 9
pas
S 8
A
cer
pt 9
pas
S 8
B

ECOLOGY AND NOTES ON SPECIES OF *THAUMALEA*

The eggs of *Thaumalea testacea* are usually laid singly, or in small clutches of 2 to 4, each being attached to the substratum by a slender film of mucus (Mandaron 1963). The ovipositing females avoid light and crawl under bryophytes or into cracks in rocks to lay the eggs (Popham 1952).

The larvae of trickle midges are characteristic of unpolluted springheads, stony streams and films of water flowing over rock surfaces. The latter include cliffs, quarry walls, rock-cuttings and brickwork, stone or concrete linings of embankments and even bridges. While no systematic studies have been done on their response to pollutants, Popham (1952) reported that they were killed by tap water and I can report that when I was rearing larvae (which I had transported from Yorkshire) in Cambridge, I ran low on my supply of water from their habitat. I topped up the remnant with freshly-caught rainwater and next day used this mixture to replace the water in the rearing tubes. All of the still-surviving larvae promptly died within a few hours!

The modes of locomotion of *Thaumalea* larvae, and their method of holding fast to a rock surface – in a film of flowing water – by means of their mouthparts and prolegs, were described by Thienemann (1910) and Popham (1952, 1961). They can move about 4–5 times faster than a typical chironomid larva can crawl. They graze on the biofilm of organic debris, especially diatoms (Leathers 1922) and other microscopic organisms. The motion of the mandibles during feeding is unusual in that they are moved from a midline position outwards rather than towards the midline, as is the case with other larval Nematocera. Their scraping action, along with that of the labrum, loosens particles from the substratum that are then retained in the collecting basket formed from the modified maxillae, labrum and hypopharynx. Internally, the pharynx has a pair of curved lamellae, which evidently facilitate the ingestion of food.

Although larvae are found in stream and trickle habitats, they are very wary of the actual running water. Essentially they are creatures that graze within the damp margins of such habitats. Thus Popham (1952) demonstrated that with *T. testacea* larvae, the youngest instars occur amongst bryophytes or in cracks in rocks, the young larvae occur in water of slight flow while the older larvae tend to avoid the flowing water by moving into the damp region on the rock, or other substratum, beyond it. Larvae generally tend to move away from the light and tend to climb up at night, but show some downward migration by day.

Mandaron (1963) reported that *T. testacea* mated in specimen tubes, in the laboratory at various times of the day, and that a male will mate successively with several females. Adult trickle midges are reported to swarm briefly around dawn (Wagner 1997c).

Published information on the distribution of the species in the British Isles has advanced little since Edwards (1929). Ashe (1987) and Ashe, O'Connor & Murray (1998) have summarised more recent records for Ireland. All of these records were based on adults. These data are supplemented by unpublished information, especially records provided by Peter Chandler, and by my own records based on larvae and adults. The recorded phenology of the larvae reflects the paucity of records. They presumably also occur at least in all of the months for which there are adult records.

Thaumalea testacea

Adults recorded March–October; larvae recorded in April and June–August. *T. testacea* seems to prefer less acid waters than the other two British species, and is characteristic of springheads on limestone and also trickles on bridges and walls enriched by the lime-rich mortar. It is not known whether this is a true preference or whether the presence of lime serves to buffer polluting acids in the rain that would otherwise inhibit the larvae. Popham (1952) provides details of the microhabitat distribution of larvae of different ages (summarised above).

Popham (1952) describes oviposition behaviour and Mandaron (1963) details the mode of oviposition (summarised above). He reported females laying 250–300 eggs in clusters of 1 to 4, but usually only one; duration of the egg stage was 5–6 days and that of first-instar larvae was 3–4 days at 16°C. Total larval duration was 58 days or more and the pupal stage was 8 days. In my own rearing experiments, I have recorded the duration of the last-instar larva as 4 days.

Taking advantage of finding *T. testacea* coexisting with *T. verralli* in the absence of *T. truncata*, the widths (breadths) of the anterior spiracles (see page 93) were measured in a sample of larvae collected from a stream at Langcliffe on 4 August 1998. Fuller details are given below under *T. verralli*. The percentage frequencies in each size-class are given in Table 2.

In August, the commonest size-class for anterior spiracles in *T. testacea* was the same (25–40 μm breadth) as in the coexisting larvae of *T. verralli* (see Table 2). But the population of the latter contained many young larvae (40.6%) in the 10-20 μm size-class whereas the population of *T. testacea*

Table 2. Size-classes of the breadth of anterior spiracles (μm) in larvae of two species of *Thaumalea* collected in the summer of 1998 from Langcliffe, near Settle, and Fountains Fell, both in north-west Yorkshire. The frequencies in each size-class are expressed as percentages of the number of larvae (N) in each sample; the highest percentages are shown in **bold** type.

Species	Locality	Month	Size-classes: N	10–20	25–40	45–60	65–80
T. verralli	Fountains	June	65	12.3	**52.3**	29.3	6.1
	Fountains	July	70	10.0	21.5	32.8	**35.7**
	Langcliffe	August	32	40.6	**59.4**	0	0
T. testacea	Langcliffe	August	29	0	**44.8**	34.5	20.7

contained none. Indeed, the age structure of the latter's population more closely resembled that of the *T. verralli* population sampled in June at more than twice the altitude, on Fountains Fell (Table 2). Only the current difficulty of distinguishing the younger larvae of *T. testacea* from those of *T. truncata* prevented a direct comparison of the age structures of all three species at the higher altitude (640 m). However, these limited data suggest that coexisting populations of the three species of trickle midges have different phenologies that are out of phase with each other.

Distribution. Scotland, Wales, the northern half of England and the south-west, with scattered localities as far apart as Sussex and Ireland.

Thaumalea truncata

Adults and larvae recorded June–August. This species seems to be restricted to more acid waters than those favoured by *T. testacea*, but otherwise typically coexists with both the other species.

In my rearing experiments, I have recorded the duration of the pupal stage as 7.5 days.

Distribution. Widely distributed in Scotland and in scattered localities in the Pennines, Cumbria and Northumberland. Elsewhere in England it is recorded from a few localities in Cornwall, Devon, Dorset, Gloucestershire, Somerset and Worcestershire, and in Wales from Glamorgan.

Thaumalea verralli

Adults recorded April–October; larvae recorded June–September. *T. verralli* seems to tolerate a greater pH range than *T. testacea* and *T. truncata*. Whenever two species have been found coexisting, *T. verralli* has always been one of them.

The Holarctic distribution of this species reflects the tolerance of its larvae to a greater temperature range than the other two species. Indeed, in Iceland it has been recorded from hot springs at 27–28°C (Andersson 1967). However, the larvae are typically found in cooler waters.

As larvae of all ages are readily distinguished from the other two species, I present some data on the frequency of larvae of different sizes (using the breadths of the anterior spiracles measured to the nearest 5 μm – see p. 93). Two populations were sampled in North Yorkshire in the summer of 1998. The first, on Fountains Fell (at 640 m altitude), was sampled in the first weeks of both June and July. The second population, a roadside stream above the school at Langcliffe village near Settle (at 200 m altitude), was sampled in the first week of August. The percentage frequencies for the different size-classes are given in Table 2.

The results serve to indicate that such simple measurements can yield information on the age structure of populations at different times and different altitudes. Furthermore, the age structure of the *T. verralli* population at Langcliffe was significantly different from that of the coexisting population of *T. testacea* larvae (see above, and Table 2).

Gravid females carry *ca.* 70 to 100 nearly-mature eggs. In my rearing experiments, I have recorded the duration of the presumed L–3 instar (see pages 93–94) as 2–5 days, the L-2 instar as 5 days, the penultimate (L–1) instar as 3–4 days (typically 4), the last instar as 4–5 days, and duration of the pupal stage as 7.5–8.5 days.

Distribution. Scattered localities throughout Ireland, Scotland, Wales and the northern half and south-west of England.

ACKNOWLEDGEMENTS

The drawings of Dixidae without scale bars (except Figs 21C and 21D) were those executed by Joan Worthington for the first edition. It was through working with Joan on the production of these figures that I not only learned much about the art of biological illustration but also that such drawings depend as much upon scientific understanding and selectivity as upon skilled draughtsmanship. The photographs in Plates 4 and 5 were taken by A. E. Ramsbottom for the first edition. The photographs in Plates 2, 3 and 7 were taken by the late Dr Kathleen Goldie-Smith. Those in Plates 6 and 10 were taken by myself, along with the SEM photographs in Plates 8 and 9, which were taken with W. M. Lee (Department of Zoology, Cambridge University) skilfully operating the SEM. Figures 9, 16C, D, 25 and 32 are by myself but were first published in the *Entomologist's Monthly Magazine*. They are reused here with the permission of the Gem Publishing Company. The remaining illustrations were specially prepared for this edition.

My work on Diptera is currently funded by the Isaac Newton Trust (Trinity College, Cambridge), to whom I am most grateful. In practice, the inadequate budgets of our Research Councils no longer allow funding of basic (alpha) taxonomy and the production of identification handbooks (Disney 1998).

I am grateful to Mark Telfer (ITE Environmental Information Centre) for the loan of unpublished BRC record cards for Dixidae. Peter Chandler kindly sent me many unpublished records for both families and also valuable specimens (especially of trickle midges) and likewise Dr Martin Drake. Ivan Perry is also thanked for allowing me to examine his collection of Thaumaleidae.

When the late Dr Kathleen Goldie-Smith took over the Dixid Recording Scheme from me she did much to advance our knowledge of the distributions of British species. Furthermore, she also carried out painstaking studies of dixid eggs and of a common virus infection of meniscus midge larvae. Her quiet, self-effacing, but dogged enthusiasm are sadly missed. Her papers and specimens have been deposited in the Cambridge University Zoology Museum.

REFERENCES

Andersson, H. (1967). Faunistic, ecological and taxonomic notes on Icelandic Diptera. *Opuscula entomologica* **32**, 101-120.

Ashe, P. (1986). A checklist of Irish Dixidae. *Bulletin of the Irish Biogeographical Society* **9**, 46-50.

Ashe, P. (1987). The families Anisopodidae and Thaumaleidae (Diptera: Nematocera) in Ireland with comments on relevant Haliday manuscripts. *Bulletin of the Irish Biogeographical Society* **10**, 107-122.

Ashe, P. & O'Connor, J. P. (1990). Further records of Irish Dixidae (Diptera) including *Dixella attica* Pandazis, new to Ireland. *Bulletin af the Biogeographical Society* **13**, 23-28.

Ashe, P., O'Connor, J. P. & Murray, D. A. (1998). *A checklist of Irish aquatic insects.* Occasional Publication of the Irish Biogeographical Society, No. 3. Dublin.

Bertrand, H. (1948). Note su la capture d'un Diptère nouveau pour la faune française. *Bulletin de la Société entomologique de France* **52**, 165-168.

Brindle, A. (1963). Taxonomic notes on the larvae of British Diptera: No. 15 – The Dixinae (Culicidae). *Entomologist* **96**, 237-243.

Christophers, R. (1960). *Aedes aegypti. The yellow fever mosquito, its life history, bionomics and structure.* Cambridge.

Colless, D. H. & McAlpine, D. K. (1991). Diptera. In *The insects of Australia,* 2nd edition, Vol. 2. CSIRO (Sponsor). Melbourne University Press **39**, 717-786.

Crosskey, R. W. (1990). *The natural history of blackflies.* John Wiley & Sons, Chichester, England.

Disney, R. H. L. (1972). Observations on sampling pre-imaginal populations of blackflies (Dipt., Simuliidae) in West Cameroon. *Bulletin of Entomological Research* **61**, 485-503.

Disney, R. H. L. (1974a). The larvae of two species of Dixidae (Dipt.) from Cameroon. *Entomologist's Monthly Magazine* **109**, 117-119 (1973).

Disney, R. H. L. (1974b). A meniscus midge new to Britain with a revised check list of the British Dixidae (Dipt.). *Entomologist's Monthly Magazine* **109**, 184-186 (1973).

Disney, R. H. L. (1975). A key to the larvae, pupae and adults of the British Dixidae (Diptera). The meniscus midges. *Scientific Publications of the Freshwater Biological Association,* No. **31**, 78 pp.

Disney, R. H. L. (1983). A synopsis of the taxonomist's tasks, with particular attention to phylogenetic cladism. *Field Studies* **5**, 841-865.

Disney, R. H. L. (1992). A meniscus midge (Dipt., Dixidae) new to Britain. *Entomologist's Monthly Magazine* **128**, 165-169.

Disney, R. H. L. (1994). From field studies to taxonomy. *Field Studies* **8**, 197-216 (1993).

Disney, R. H. L. (1998). Rescue plan needed for taxonomy. *Nature* **394**, 120.

Edwards, F. W. (1920). The British Chaoborinae and Dixinae. *Entomologist's Monthly Magazine* **56**, 264-270.

Edwards, F. W. (1929). A revision of the Thaumaleidae (Dipt.). *Zoologischen Anzeiger* **82**, 121-142.

Elliott, J. M. (1967). Invertebrate drift in a Dartmoor stream. *Archiv für Hydrobiologie* **63**, 202-237.

Elliott, J. M. & Minshall, G. W. (1968). The invertebrate drift in the River Duddon, English Lake District. *Oikos* **19**, 39-52.

Elliott, J. M. & Tullett, P. A. (1977). The downstream drifting of larvae of *Dixa* (Diptera: Dixidae) in two stony streams. *Freshwater Biology* **7**, 403-407.

Fowler, J. A. (1984a). Meniscus midges (Dipt., Dixidae) new to Leicestershire and Rutland. *Entomologist's Monthly Magazine* **120**, 113-114.

Fowler, J. A. (1984b). The status of *Dixella martinii* Peus (Dipt., Dixidae) in Shetland. *Entomologist's Monthly Magazine* **120**, 114.

Fowler, J. A. (1987a). A meniscus midge (Diptera: Dixidae) on Fair Isle. *Fair Isle Bird Observatory Report* **39** (1986), 64.

Fowler, J. A. (1987b). A meniscus midge (Dipt., Dixidae) new to Wales. *Entomologist's Monthly Magazine* **123**, 178.

Fowler, J. A. (1989). Meniscus midges (Dipt., Dixidae) on Ynys Enlli (Bardsey Is.). *Entomologist's Monthly Magazine* **125**, 118.

Fowler, J. A., Withers, I. D. & Dewhurst, F. (1997). Meniscus midges (Diptera: Dixidae) as indicators of surfactant pollutants. *Entomologist* **116**, 24-27.

Freeman, P. (1950). Family Culicidae. Subfamilies Dixinae and Chaoborinae. *Handbooks for the identification of British insects* **9 (2)**, 97-101.

Frizzi, G. & Contini, C. (1962). Studio introduttivo citogenetico su alcune specie di Dixidae della Sardegna. *Bolletino di Zoologia* **29**, 621-633.

Frizzi, G., Contini, C. & Mameli, M. (1966). Ulteriori ricerche citogenetiche sui Dixidae della Sardegna. *Atti, Associazone Genetica Italiana, Pavia* **11**, 286-291.

Goldie-Smith, E. K. (1987). Virus infections of larval Dixidae and other Culicoidea (Diptera: Nematocera). *Entomologist's Gazette* **38**, 137-139.

Goldie-Smith, E. K. (1989a). Laboratory rearing and egg development in two species of *Dixella* (Diptera: Dixidae). *Entomologist's Gazette* **40**, 53-65.

Goldie-Smith, E. K. (1989b). The eggs of *Dixella aestivalis* Meigen, and brief comparisons with eggs of three other species of *Dixella* (Dipt., Dixidae). *Entomologist's Monthly Magazine* **125**, 107-117.

Goldie-Smith, E. K. (1990a). The eggs of *Dixella obscura* Loew, *D. attica* Pandazis and *Dixa nubilipennis* Curtis (Diptera: Dixidae). *Dipterists Digest* **3**, 2-7 (1989).

Goldie-Smith, E. K. (1990b). Distribution maps for Dixidae in Great Britain and Ireland. *Dipterists Digest* **3**, 8-26 (1989).

Goldie-Smith, E. K. (1993). Observations of the behaviour and immature stages of *Dixella graeca* Pandazis (Diptera: Dixidae). *Dipterists Digest* **13**, 2-5.

Goldie-Smith, E. K. & Thorpe, J. R. (1991). Eggs of British meniscus midges (Diptera: Dixidae) observed by scanning electron miscoscopy. *Freshwater Forum* **1**, 215-224.

Guthrie, M. (1989). *Animals of the surface film.* Naturalists' Handbook No. **12**, 87 pp. Richmond Publishing Co. Ltd., Slough, England.

Johannsen, O. A. (1934). Aquatic Diptera Part 1. Nematocera, exclusive of Chironomidae and Ceratopogonidae. *Memoirs of the Cornell University*

Agricultural Experimental Station **164**, 1-72.

Jones, R. K. H. (1965). *Ecology of the Hydracarina.* Unpublished PhD thesis, Leicester.

Kloet, G. S. & Hincks, W. D. (1945). A *check list of British insects.* Kloet & Hincks, Stockport.

Kloet, G. S. & Hincks, W. D. (1976). A check list of British insects. 2nd edition. Part 5. Diptera and Siphonaptera. *Handbooks for the identification of British Insects* **11 (5)**, 1-139 (1975).

Leathers, A. L. (1922). Ecological study of aquatic midges and some related insects with special reference to feeding habits. *Bulletin of the Bureau of Fisheries, Washington* **38**, 1-61.

Lindner, E. (1930). 3. Thaumaleidae (Orphnephilidae). In *Die Fliegen der palaearktischen Region* (ed. E. Lindner) **2** (1), 1-16. E. Schweizerbart'sche Verrlagsouchhandlung, Stuttgart, Germany.

Mandaron, P. (1963). Accouplement, ponte et premiere stade larvaire de *Thaumalea testacea* Ruthé (Diptères Nematocères). *Travaux du Laboratoire d'Hydrobiologie et Pisciculture de l'Universite de Grenoble* **54-55**, 97-107.

Miall, L. C. (1895). *The natural history of aquatic insects.* Macmillan, London.

Miller, B. R., Crabtree, M. B. & Savage, H. M. (1997). Phylogenetic relationships of the Culicomorpha inferred from 18s and 5.8s ribosomal DNA sequences (Diptera: Nematocera). *Insect Molecular Biology* **6**, 105-114.

Morgan, N. C. & Waddell, A. B. (1961). Diurnal variation in the emergence of some aquatic insects. *Transactions of the Royal Entomological Society of London* **113**, 123-137.

Nicholson, P. A. (1978). Some observations on *Dixella autumnalis* Meigen (Dipt., Dixidae). *Entomologist's Monthly Magazine* **113**, 62 (1977).

Nicholson, P. A. (1979). Observations on *Dixella aestivalis* Meigen (Dipt., Dixidae). *Entomologist's Monthly Magazine* **114**, 156 (1978).

Nielsen, P., Ringdahl, O. & Tuxen, S. L. (1954). Diptera I (exclusive of Ceratopogonidae and Chironomidae). *Zoology of Iceland* **3 (48a)**. Ejnar Munksgaard, Copenhagen & Reykjavik.

Nowell, W. R. (1951). The dipterous family Dixidae in western North America (Insecta: Diptera). *Microentomology* **16**, 187-270.

Oldroyd, H. (1970). Introduction and key to families (3rd edition). *Handbooks for the identification of British insects* **9 (1)**, 1-104.

Pawlowski, J., Szadziewski, R., Knieciak, D., Fahrni, J. & Bittar, G. (1996). Phylogeny of the infraorder Culicomorpha (Diptera: Nematocera) based on 28S RNA gene sequences. *Systematic Entomology* **21**, 167-178.

Peach, W. J. (1984). *Ecology and life history of meniscus midges (Diptera: Dixidae).* Unpublished BSc thesis. Leicester Polytechnic.

Peach, W. J. & Fowler, J. A. (1986). Life cycle and laboratory culture of *Dixella autumnalis* Meigen (Dipt., Dixidae). *Entomologist's Monthly Magazine* **122**, 59-62.

Peach, W. J. & Kinsler, R. A. (1988). Seasonal changes in sex ratio in *Dixa nebulosa* Meigen (Dipt., Dixidae). *Entomologist's Monthly Magazine* **124**, 83-86.

Peters, T. M. & Adamski, D. (1982). A description of the larva of *Dixella nova* (Walker) (Diptera: Dixidae). *Proceedings of the Entomological Society, of Washington* **84**, 521-528.

Peters, T. M. & Cook, E. F. (1966). The Nearctic Dixidae (Diptera). *Miscellaneous Publications of the Entomological Society of America* **5**, 233-278.

Popham, E. J. (1952). Some preliminary notes on the fauna hygropetrica of Lancashire and the Isle of Man. *Journal of the Society for British Entomology* **4**, 59-63.

Popham, E. J. (1961). *Some aspects of life in fresh water.* 2nd Edition. London.

Rao, P. N. & Rai, K. S. (1990). Genome evolution in the mosquitoes and other closely related members of the superfamily Culicoidea. *Heriditas* **113**, 139-144.

Roper, P. (1962). Some notes on the Dixinae (Diptera: Culicidae) of East Sussex. *Entomologist's Record* **74**, 21-23.

Saunders, L. G. (1923). On the larva, pupa, and systematic position of *Orphnephila testacea,* Macq. (Diptera, Nematocera). *Annals and Magazine of Natural History (9)* **11**, 631-640.

Schmid, F. (1951). Notes sur quelques Thaumaléides suisses et espagnols (Diptera, Nematocera). *Bulletin de l'Institut Royal des Sciences naturelles de Belgique* **27 (40)**, 1-6.

Sicart, M. (1959). Dixinae de sud de la France. *Bulletin de la Societé d'Histoire Naturelle de Toulous* **94**, 288-324.

Sinclair, B. J. (1996). A review of the Thaumaleidae (Diptera: Culicomorpha) of eastern North America, including a redefinition of the genus *Androposopa* Mik. *Entomologica scandinavica* **27**, 361-376.

Stechmann, D.-H. (1977). Zur Phänologie und zum Wirtsspektrum einiger an Zygopteren (Odonata) und Nematoceren (Diptera) ektoparasitischauftretenden *Arrenurus*-Arten (Hydrachnellae, Acari). *Zeitschrift für angewandte Entomologie* **82**, 349-355.

Stechmann, D.-H. (1978). Eiablage, Parasitismus und postparasitische

Entwicklung von *Arrenurus*-Arten (Hydrachnellae, Acari). *Zeitschrift für Parasitentunde* **57**, 169-188.

Stone, A. & Peterson, B. V. (1981). Thaumaleidae. In *Manual of Nearctic Diptera* (eds J. F. McAlpine, B. V. Peterson, G. E. Shewell, H. J. Teskey, J. R. Vockereth & D. M. Wood). Vol. 1, Monograph 27; **26**, 351-353. Research Branch, Agriculture Canada, Ottawa.

Theabald, F. V. (1892). An *account of British flies (Diptera).* Vol. 1. Elliot Stock, London.

Thienemann, A. (1910). *Orphnephila testacea* Macq. ein Beitrag zur Kenntnis der Fauna hygropetrica. *Annales de Biologie Lacustre, Bruxelles* **4 (1909)**, 53-87.

Thomas, A. B. G. (1979). Diptères torrenticoles peu connus. VI. Les Dixidae du sud-ouest de la France (Nematocera) *(Dixa puberula* Loew, 1849. Écologie, microhabitat et intérêt pratique pour le dépistage des pollutions per les stations touristiques de montagne). *Bulletin de la Societé d'Histoire Naturelle de Toulouse* **115**, 242-268.

Tuxen, S. L. (1969). Nomenclature and homology of genitalia in insects. *Memorie della Società Entontomologica Italiana* **48**, 6-18.

Upton, M. S. (1993). Aqueous gum-chloral slide mounting media: an historical review. *Bulletin of Entomological Research* **83**, 267-274.

Vaillant, F. (1959). Quelques Dixidae paléarctiques et les habitats de leurs larves (Dipt.). *Bulletin de la Societé Entomologie de France* **64**, 178-186.

Vaillant, F. (1965). Quelques Dixidae paléarctiques nouveaux ou mal connus (Diptera). *Annales de la Societé de Entomologique de France (N.S.)* **1**, 789-795.

Vaillant, F. (1969). Les Diptères Dixidae des Pyrénées, des Alpes et des Carpates. *Annales de Limnologie* **5**, 73- 84.

Vaillant, F. (1977). Les Diptères Thaumaleidae d'Europe. *Annales de la Societé de Entomologique de France (N.S.)* **13**, 695-710.

Vaillant, F. (1981). Some Diptera Thaumaleidae from Europe. *Aquatic Insects* **3**, 129-146.

Wagner, R. (1980). Die Dipterenemergenz am Breitenbach (1969-1973). *Spixania* **3**, 167-177.

Wagner, R. (1997a). Diptera Thaumaleidae. In *Aquatic insects of North Europe. Vol. 2. Odonata-Diptera* (ed. E. Nilsson), pp. 187-191. Apollo Books, Stenstrup, Denmark.

Wagner, R. (1997b). Family Dixidae. In *Contributions to a manual of Palaearctic Diptera. Vol. 2. Nematocera and lower Brachycera* (eds L. Papp & B. Darvas), **2.16**, 299-303. Science Herald, Budapest.

Wagner, R. (1997c). Family Thaumaleidae. In *Contributions to a manual of*

Palaearctic Dipteral Vol. 2. Nematocera and lower Brachycera (eds L. Papp & B. Darvas) **2.19**, 325-329. Science Herald, Budapest.

Williams, L. R. & Fowler, J. A. (1986). *Dixa attica* Pandazis (Dipt., Dixidae) in a restored pond in Middlesex. *Entomologist's Monthly Magazine* **122**, 258.

Wright, M. J. (1901). The resistance of the larval mosquito to cold. *British Medical Journal* **1**, 882-883.